Ergebnisse der Mathematik und ihrer Grenzgebiete Band 87

Christian Berg · Gunnar Forst

Potential Theory on Locally Compact Abelian Groups

Springer-Verlag
Berlin Heidelberg New York 1975

Christian Berg, Gunnar Forst
University of København, 2100 København, Denmark

ISBN-13:978-3-642-66130-3 e-ISBN-13:978-3-642-66128-0
DOI: 10.1007/978-3-642-66128-0

Library of Congress Cataloging in Publication Data
Berg, Christian, 1944–
Potential theory on locally compact Abelian groups.
(Ergebnisse der Mathematik und ihrer Grenzgebiete; Bd. 87)
"The book grew out of lectures given at the University of Copenhagen 1973–74."
Bibliography: p.
Includes index.
1. Potential, Theory of. 2. Locally compact Abelian groups. I. Forst, Gunnar, 1946– II. Title.
III. Series.
QA404.7.B47 515'.7 75-14051
ISBN-13:978-3-642-66130-3

AMS Subject Classification (1975): 31C05, 31C15, 43A25, 43A35, 43A40, 47D05, 60B15, 60J15, 60J45, 60J65

Preface

Classical potential theory can be roughly characterized as the study of Newtonian potentials and the Laplace operator on the Euclidean space $\mathbb{R}^3$. It was discovered around 1930 that there is a profound connection between classical potential theory and the theory of Brownian motion in $\mathbb{R}^3$. The Brownian motion is determined by its semigroup of transition probabilities, the Brownian semigroup, and the connection between classical potential theory and the theory of Brownian motion can be described analytically in the following way: The Laplace operator is the infinitesimal generator for the Brownian semigroup and the Newtonian potential kernel is the "integral" of the Brownian semigroup with respect to time.

This connection between classical potential theory and the theory of Brownian motion led Hunt (cf. Hunt [2]) to consider general "potential theories" defined in terms of certain stochastic processes or equivalently in terms of certain semigroups of operators on spaces of functions.

The purpose of the present exposition is to study such general potential theories where the following aspects of classical potential theory are preserved:

(i) The theory is defined on a *locally compact abelian group.*

(ii) The theory is *translation invariant* in the sense that any translate of a potential or a harmonic function is again a potential, respectively a harmonic function; this property of classical potential theory can also be expressed by saying that the Laplace operator is a differential operator with *constant coefficients.*

(iii) The theory is *submarkovian;* this is no restriction in relation to the theory of Hunt, and it corresponds to the property in classical potential theory, that the non-negative constant functions on $\mathbb{R}^3$ are superharmonic.

(iv) The theory is *transient,* which corresponds to the fact in classical potential theory that the integral of the Brownian semigroup (on $\mathbb{R}^3$) with respect to time converges and defines a positive measure.

The basic datum for such a "potential theory" is a family of positive bounded measures $(\mu_t)_{t>0}$, called a *convolution semigroup,* on a locally compact abelian group G, with the property that the integral $\int_0^\infty \mu_t dt$ converges and defines a positive measure κ, called the *potential kernel.* The potential theory then consists in a study of the convolution semigroup $(\mu_t)_{t>0}$, the potential kernel κ and concepts derived from these notions.

Many of the concepts treated in such a potential theory have probabilistic interpretations, but these will play only a minor role in the present treatment where (almost) only "analytical" methods, in particular from Fourier analysis, are applied.

The book is divided in three chapters.

The first chapter is of preparatory nature, containing mostly well-known facts from harmonic analysis. The theory of Fourier transformation of positive definite measures, which is not easily accessible in the literature, will be presented in a fairly detailed way (§ 4).

The second chapter is devoted to the study of convolution semigroups and the strongly continuous contraction semigroups they induce. The first main result is the characterization of convolution semigroups in terms of continuous negative definite functions on the dual group (§ 8). In an important special case, the convolution semigroups "on" the positive half-axis, this characterization may be interpreted via the set of Bernstein functions (§ 9). Then many of the "classical" examples of convolution semigroups and their corresponding continuous negative definite functions on the group $\mathbb{R}^n$ are discussed (§ 10). We start the study of the induced contraction semigroups by fixing the notation in the situation of an abstract contraction semigroup (§ 11), and the theory is then specialized to translation invariant contraction semigroups (§ 12).

The third chapter contains the "potential theory". First the transient convolution semigroups are introduced and the study of the potential kernel is initiated; in particular some results on the behaviour at infinity are given (§ 13). The set of all potential kernels for transient convolution semigroups is then studied using the special case of convolution semigroups on the positive half-axis, and some interesting convex cones of potential kernels are exhibited (§ 14). In the further description of the potential kernel it is shown that it satisfies "many" potential theoretic principles (§ 15, § 16), and this is based on a discussion of excessive and invariant measures (§ 16). The potential kernel can be characterized as a "perfect" kernel (§ 17), and in the last paragraph (§ 18) the infinitesimal generator is studied, and we consider in particular convolution semigroups for which the infinitesimal generator is a local operator.

Numerous concrete examples and exercises are inserted in the text. The exercises fall into two categories, one consisting of example-like exercises and the other consisting of exercises of a more theoretical nature, where the reasoning mainly follows the lines indicated in the surrounding text. Occasionally the result of such an exercise will be used in a proof.

The reason for writing the book was to give a self-contained introduction to convolution semigroups and their potential theory in the framework of locally compact abelian groups.

Most of the material is well-known to specialists, but some of the subjects treated are not easily accessible in the existing literature. The bibliographical comments at the ends of the paragraphs indicate our sources, and although we did not aim at completeness with these comments, it is our hope that we have been fairly honest. In any case it has not been the intention to give anyone less than full credit, and we apologize in advance for all omissions and errors.

The book grew out of lectures given at the University of Copenhagen 1973–74, and we want to express our gratitude to the students, who made it a pleasure to give these lectures, and to Heinz Bauer for his interest in making the material into this book. Finally we want to thank Ilse Dyrby for her efficient help with the typing of the manuscript.

October 1974

Christian Berg
Gunnar Forst

Contents

Chapter I. Harmonic Analysis 1

§ 1. Notation and Preliminaries 1
§ 2. Some Basic Results From Harmonic Analysis 8
§ 3. Positive Definite Functions 11
§ 4. Fourier Transformation of Positive Definite Measures 17
§ 5. Positive Definite Functions on ℝ 26
§ 6. Periodicity 30

Chapter II. Negative Definite Functions and Semigroups 39

§ 7. Negative Definite Functions 39
§ 8. Convolution Semigroups 48
§ 9. Completely Monotone Functions and Bernstein Functions 61
§ 10. Examples of Negative Definite Functions and Convolution Semigroups 72
§ 11. Contraction Semigroups 76
§ 12. Translation Invariant Contraction Semigroups 85

Chapter III. Potential Theory for Transient Convolution Semigroups . . . 97

§ 13. Transient Convolution Semigroups 97
§ 14. Transient Convolution Semigroups on the Half-Axis and Integrals of Convolution Semigroups 121
§ 15. Convergence Lemmas and Potential Theoretic Principles 136
§ 16. Excessive Measures 146
§ 17. Fundamental Families Associated With Potential Kernels 160
§ 18. The Lévy Measure for a Convolution Semigroup 171

Bibliography 191

Symbols 195

General Index 196

Chapter I. Harmonic Analysis

§ 1. Notation and Preliminaries

In this first paragraph the basic notation and terminology will be established, and we shall in particular discuss various sets of functions and measures (together with their topologies) on a locally compact abelian group (for short just *LCA-group*) G. The reader is a assumed to be familiar with convolution of measures and we refer to the book of Hewitt and Ross [1] for a detailed exposition. In connection with convolution of unbounded measures we introduce shift-bounded measures and measures tending to zero at infinity.

1.1. Let G be a LCA-group, where the composition is written additively and the neutral element is denoted 0.

We denote by $C(G)$ the set of continuous complex functions on G. By $C_c(G)$, resp. $C_0(G)$, resp. $C_b(G)$, we denote the set of functions from $C(G)$ which have compact support, resp. which tend to zero at infinity, resp. which are bounded.

All these spaces will be endowed with their canonical topologies, i.e. $C(G)$ with the topology of compact convergence, $C_c(G)$ with the inductive limit topology and $C_0(G)$ and $C_b(G)$ with the topology of uniform convergence.

The dual space of $C_c(G)$ is the space of Radon measures on G, denoted $M(G)$. The dual space of $C_0(G)$ is identified with the subset $M_b(G)$ of $M(G)$ consisting of the bounded Radon measures on G. By $M_c(G)$ we denote the set of Radon measures on G with compact support.

All measures under consideration in the sequel will tacitly be assumed to be Radon measures.

If $A(G)$ is a set of functions or measures on G, $A^+(G)$ denotes the set of positive elements in $A(G)$.

The set of probability measures on G is denoted $M_1^+(G)$. The one-point measure (or Dirac measure) at $a \in G$ is denoted ε_a.

1.2. For a function f on G, $\check{f}$ and $\tilde{f}$ denote the functions defined by

$$\check{f}(x) = f(-x) \quad \text{and} \quad \tilde{f}(x) = \overline{\check{f}}(x) = \overline{f(-x)} \quad \text{for } x \in G.$$

For a measure μ on G, $\check{\mu}$ and $\tilde{\mu}$ denote the measures defined by

$$\langle \check{\mu}, f \rangle = \langle \mu, \check{f} \rangle \quad \text{and} \quad \langle \tilde{\mu}, f \rangle = \overline{\langle \mu, \tilde{f} \rangle} \quad \text{for } f \in C_c(G).$$

For $a \in G$ we define the *translation* τ_a: $G \to G$ by $\tau_a x = a + x$. For a function f on G, $\tau_a f$ denotes the function

$$\tau_a f(x) = f(x - a) \quad \text{for } x \in G,$$

and for a measure μ on G, $\tau_a\mu$ denotes the measure defined by

$$\langle \tau_a \mu, f\rangle = \langle \mu, \tau_{-a} f\rangle \quad \text{for } f \in C_c(G).$$

1.3. The weak topology $\sigma(M(G), C_c(G))$ is called the *vague topology*. A net $(\mu_\alpha)_{\alpha\in A}$ of elements from $M(G)$ converges vaguely (i.e. in the vague topology) to an element $\mu \in M(G)$ if and only if

$$\lim_A \langle \mu_\alpha, f\rangle = \langle \mu, f\rangle \quad \text{for all } f \in C_c(G).$$

The vector spaces $M_b(G)$ and $C_b(G)$ form a dual pair, and the weak topology $\sigma(M_b(G), C_b(G))$ will be called the *Bernoulli topology*. A net $(\mu_\alpha)_{\alpha\in A}$ of elements from $M_b(G)$ converges to an element $\mu \in M_b(G)$ in the Bernoulli topology if and only if

$$\lim_A \langle \mu_\alpha, f\rangle = \langle \mu, f\rangle \quad \text{for all } f \in C_b(G).$$

For positive bounded measures we have the following relation between the Bernoulli topology and the vague topology.

1.4. Proposition. *A net $(\mu_\alpha)_{\alpha\in A}$ on $M_b^+(G)$ converges to $\mu \in M_b^+(G)$ in the Bernoulli topology if and only if $(\mu_\alpha)_{\alpha\in A}$ converges to μ vaguely and furthermore*

$$\lim_A \mu_\alpha(G) = \mu(G).$$

Proof. The "only if"-part is clear. Suppose that μ_α tends to μ vaguely and that $\mu_\alpha(G)$ tends to $\mu(G)$ and let $f \in C_b(G)$ and $\varepsilon > 0$ be given.

Since μ is a bounded measure there exists a compact subset $K \subseteq G$ such that $\mu(G \setminus K) < \varepsilon$. Let us choose a function $h \in C_c(G)$ such that $0 \leqq h \leqq 1$ and $h = 1$ on K.

By the assumptions we have

$$\lim_A \langle \mu_\alpha, 1-h\rangle = \langle \mu, 1-h\rangle \leqq \mu(G\setminus K) < \varepsilon,$$

and

$$\lim_A \langle \mu_\alpha, f h\rangle = \langle \mu, f h\rangle,$$

and there exists therefore $\alpha_0 \in A$ such that for $\alpha \in A$ and $\alpha \geqq \alpha_0$

$$\langle \mu_\alpha, 1-h\rangle < \varepsilon \quad \text{and} \quad |\langle \mu, f h\rangle - \langle \mu_\alpha, f h\rangle| < \varepsilon.$$

For $\alpha \in A$ and $\alpha \geqq \alpha_0$ we then have

$$\begin{aligned} &|\langle \mu, f\rangle - \langle \mu_\alpha, f\rangle| \\ &\leqq |\langle \mu, f h\rangle - \langle \mu_\alpha, f h\rangle| + |\langle \mu, f(1-h)\rangle| + |\langle \mu_\alpha, f(1-h)\rangle| \\ &\leqq \varepsilon(1 + 2\|f\|_\infty), \end{aligned}$$

which shows that $\lim_A \mu_\alpha = \mu$ in the Bernoulli topology. □

1.5. Once and for all we choose and fix a Haar measure ω_G on G. In formulas we often write dx instead of ω_G. "Almost everywhere on G" will mean almost everywhere with respect to the Haar measure on G, and it is abbreviated a.e. The L^p-spaces with respect to ω_G are denoted $L^p(G)$ for $p \in [1, \infty]$.

For a locally integrable Borel function f on G we denote by $f\omega_G$ the measure defined by

$$\langle f\omega_G, g\rangle = \int f(x) g(x) dx \quad \text{for } g \in C_c(G).$$

Occasionally we will not distinguish between the function f and the measure $f\omega_G$.

1.6. Definition. An *approximate unit* on G is a net $(\varphi_V)_{V\in\check{V}}$ of functions $\varphi_V \in C_c^+(G)$, where $\check{V}$ is a basis for the filter of neighbourhoods of the neutral element in G and such that for every $V \in \check{V}$

$$\operatorname{supp}(\varphi_V) \subseteq V \quad \text{and} \quad \langle \omega_G, \varphi_V\rangle = 1.$$

For an approximate unit $(\varphi_V)_{V\in\check{V}}$ the net $(\varphi_V \omega_G)_{V\in\check{V}}$ of probability measures converges to ε_0 in the Bernoulli topology.

1.7. It is well-known that there is defined a *convolution* $\mu * \nu$ between certain measures or functions μ and ν on G.

In particular convolution is a well-defined operation on $M_b(G)$, and with this operation $M_b(G)$ is a commutative Banach algebra.

Furthermore $L^1(G)$ is a commutative Banach algebra under convolution, and the mapping $f \mapsto f\omega_G$ is an isometric injection of $L^1(G)$ onto a closed ideal in $M_b(G)$.

It is perhaps less standard to define the convolution for non-bounded positive measures.

1.8. Definition. Let μ and ν be positive measures on G. We will say that the *convolution of μ and ν exists*, if

$$\int_{G\times G} f(x+y) d\mu\otimes\nu(x, y) < \infty \quad \text{for all } f \in C_c^+(G).$$

If the convolution of μ and ν exists, the mapping

$$f \mapsto \int f(x+y) d\mu\otimes\nu(x, y)$$

of $C_c(G)$ into $\mathbb{C}$ defines a positive measure on G called the *convolution* of μ and ν and denoted $\mu * \nu$.

If the convolution of μ and ν exists so does the convolution of ν and μ and $\mu * \nu = \nu * \mu$. For all $f \in C_c(G)$ we have

$$\langle \mu * \nu, f\rangle = \langle \mu, \check{\nu} * f\rangle = \langle \nu, \check{\mu} * f\rangle. \tag{1}$$

The convolution is associative in the following sense: Let μ, ν and τ be positive measures. If the convolutions $\mu*(\nu*\tau)$ and $\mu*\nu$ exist, then $(\mu*\nu)*\tau$ exists and

$$\mu*(\nu*\tau)=(\mu*\nu)*\tau. \tag{2}$$

For a positive measure μ we denote by $D^+(\mu)$ the set of positive measures ν for which $\mu*\nu$ exists. Clearly $D^+(\mu)$ is a convex cone containing the positive measures with compact support. Furthermore, if $\nu\in D^+(\mu)$ and if τ is a positive measure such that $\tau\leqq\nu$, then $\tau\in D^+(\mu)$.

In the following shall we also use convolution of certain complex measures. We say that the convolution of two measures μ and ν exists, if the convolution of $|\mu|$ and $|\nu|$ exists, and the convolution of μ and ν is then the measure, denoted $\mu*\nu$, which is defined by

$$\langle\mu*\nu,f\rangle=\int f(x+y)\,d\mu\otimes\nu(x,y) \quad \text{for} \quad f\in C_c(G).$$

(That this expression defines a measure can be seen e.g. by writing μ and ν as linear combinations of positive measures).

1.9. Lemma. *Let μ be a non-zero positive measure on G. For every compact subset $K\subseteq G$ there exists a function $f\in C_c^+(G)$ such that*

$$\mu*f(x)\geqq 1 \quad \textit{for all } x\in K.$$

The only measure $\nu\in D^+(\mu)$ for which $\mu*\nu=0$ is the zero measure.

Proof. The exists a function $g\in C_c^+(G)$ such that $\langle\mu,g\rangle>0$. For $x\in G$ we then have

$$\mu*(\tau_x\check{g})(x)>0.$$

By a simple compactness argument there exist finitely many points $x_1,\ldots,x_n$ in G such that the function

$$f=\sum_{i=1}^{n}\tau_{x_i}\check{g}$$

satisfies $\mu*f(x)>0$ for all $x\in K$. A positive multiple of f satisfies the condition of the lemma.

Suppose that $\nu\in D^+(\mu)$ verifies $\mu*\nu=0$. For every compact set $K\subseteq G$ there exists $f\in C_c^+(G)$ such that $\check{\mu}*f(x)\geqq 1$ for $x\in K$, and hence

$$\nu(K)\leqq\langle\nu,\check{\mu}*f\rangle=\langle\mu*\nu,f\rangle=0,$$

and it follows that $\nu=0$. □

1.10. Exercise. Let μ, ν and τ be positive measures on G and suppose that $\mu\neq 0$ and that the convolutions of μ and ν and of $\mu*\nu$ and τ exist.

Then the convolutions of ν and τ and of μ and $\nu*\tau$ exist, and we have

$$\mu*(\nu*\tau)=(\mu*\nu)*\tau.$$

1.11. Definition. A measure μ on G is called (*vaguely*) *shift-bounded* if the set $\{\tau_x\mu \mid x\in G\}$ of translates of μ is vaguely bounded.

If μ is shift-bounded then $\check{\mu}$ and $\tilde{\mu}$ are shift-bounded.
A measure μ on G is shift-bounded if and only if $\mu*f\in C_b(G)$ for all $f\in C_c(G)$.

1.12. Proposition. *Let μ be a shift-bounded measure on G. Then the linear mapping $f\mapsto\mu*f$ of $C_c(G)$ into $C_b(G)$ is continuous, and the absolute value $|\mu|$ of μ is a shift-bounded measure.*

Proof. Let K be a compact subset of G and denote by $C_c(G,K)$ the space of functions $f\in C_c(G)$ with $\operatorname{supp}(f)\subseteq K$, which is a Banach space under the uniform norm. It follows by the closed graph theorem that the mapping $f\mapsto\mu*f$ of $C_c(G,K)$ into $C_b(G)$ is continuous, i.e. there exists a constant $C_K>0$ such that

$$\|\mu*f\|_\infty\leqq C_K\|f\|_\infty \quad \text{for } f\in C_c(G,K).$$

The topology on $C_c(G)$ being the inductive limit topology of the spaces $C_c(G,K)$, where K varies through the compact subsets of G, the continuity assertion of the proposition follows.
Suppose now that $f\in C_c^+(G,K)$. For $g\in C_c(G)$ such that $|g|\leqq f$ we have

$$\|\mu*g\|_\infty\leqq C_K\|g\|_\infty\leqq C_K\|f\|_\infty,$$

and in particular for all $x\in G$

$$\begin{aligned}|\mu|*f(x)&=\langle|\mu|,\tau_x(\check{f})\rangle=\sup_{|g|\leqq f}|\langle\mu,\tau_x(\check{g})\rangle|\\&=\sup_{|g|\leqq f}|\mu*g(x)|\leqq C_K\|f\|_\infty,\end{aligned}$$

and it follows that $|\mu|$ is shift-bounded. □

1.13. Proposition. *The convolution of a shift-bounded measure μ and a bounded measure ν exists.*

Proof. For $f\in C_c^+(G)$ we have

$$\int\left(\int f(x+y)\,d|\mu|(x)\right)d|\nu|(y)=\int|\check{\mu}|*f(y)\,d|\nu|(y)<\infty,$$

because $|\check{\mu}|*f$ is a bounded function by Proposition 1.12. □

1.14. Definition. A measure μ on G is said to *tend to zero at infinity* if $\mu*f\in C_0(G)$ for all $f\in C_c(G)$.

If μ tends to zero at infinity then so do $\check{\mu}$ and $\tilde{\mu}$, and μ is shift-bounded.

Every bounded measure tends to zero at infinity. In particular, if f is a continuous and integrable function, the measure $f\omega_G$ tends to zero at infinity. However, for such a function one does not necessarily have $\lim_{x\to\infty} f(x)=0$.

Any measure on a compact group tends to zero at infinity.

The Haar measure ω_G of G is shift-bounded, and it tends to zero at infinity if and only if G is compact.

1.15. Remark. Let μ be a measure which tends to zero at infinity. Then the linear mapping $f \mapsto \mu * f$ of $C_c(G)$ into $C_0(G)$ is continuous. This is proved like the analogous result in Proposition 1.12.

The absolute value $|\mu|$ of μ need not tend to zero at infinity as the following example on $G=\mathbb{R}$ shows.

Let f: $\mathbb{R} \to \mathbb{R}$ be given by:

On the interval $[n, n+1[$, $n=0, 1, 2, \ldots$, we put $f(x)=(-1)^k$ for

$$x \in \left[n+\frac{k}{2^n}, n+\frac{k+1}{2^n}\right[, \qquad k=0, 1, \ldots, 2^n-1,$$

and for $x<0$ we put $f(x)=0$.

Then the measure $\mu=f(x)dx$ tends to zero at infinity but $|\mu|=1_{[0,\infty[}(x)dx$ does not tend to zero at infinity.

1.16. Proposition. *Let μ be a measure on G and let H be a closed subgroup of G such that* $\operatorname{supp}(\mu)\subseteq H$. *Then μ is shift-bounded (resp. tends to zero at infinity) considered as a measure on H, if and only if μ is shift-bounded (resp. tends to zero at infinity) considered as a measure on G.*

Proof. The "if-part" follows immediately from the fact that every function $f \in C_c(H)$ is the restriction to H of some function from $C_c(G)$.

For the "only if-part" we first suppose that μ is shift-bounded on H. Let $f \in C_c(G)$ be given and let $K=\operatorname{supp}(f)$. The set of restrictions to H

$$A=\{(\tau_{-k}f)|H \mid k \in K\}$$

is compact in $C_c(H)$, so by Proposition 1.12 the set $\mu * A$ is compact in $C_b(H)$ and there exists consequently a number $c>0$ such that

$$\left|\int f(h+k-y)\,d\mu(y)\right| \leqq c \quad \text{for } h \in H \text{ and } k \in K.$$

Since $\mu * f(x)=0$ for $x \notin H+K$, it follows that $\mu * f$ is bounded on G.

Suppose next that μ tends to zero at infinity on H. With f, K and A as above we get that $\mu * A$ is compact in $C_0(H)$ and hence equicontinuous at infinity. For given $\varepsilon>0$ there exists a compact set $L \subseteq H$ such that

$$|\mu * ((\tau_{-k}f)\,|\,H)(h)| \leqq \varepsilon \quad \text{for } h \in H \setminus L \text{ and } k \in K,$$

i.e.

$$\left|\int f(h+k-y)\,d\mu(y)\right| \leqq \varepsilon \quad \text{for } h \in H \setminus L \text{ and } k \in K.$$

It follows easily that

$$|\mu * f(x)| \leqq \varepsilon \quad \text{for } x \in G \setminus (K+L),$$

and hence $\mu * f \in C_0(G)$. □

1.17. Exercise. A positive measure μ on G is shift-bounded (resp. tends to zero at infinity) if and only if for all compact subsets $K \subseteq G$ the function $x \mapsto \mu(x+K)$ is bounded (resp. tends to zero at infinity).

1.18. Proposition. *Let E be one of the following spaces $C(G)$, $C_c(G)$, $C_0(G)$ or $L^p(G)$ for $p \in [1, \infty[$, equipped with their usual topologies, or let E be $M(G)$ with the vague topology.*

For an approximate unit $(\varphi_V)_{V \in \check{V}}$ on G and an element $f \in E$ we have $\varphi_V * f \in E$ for all $V \in \check{V}$, and the net $(\varphi_V * f)_{V \in \check{V}}$ converges to f in the topology of E.

The proof is left as an exercise.

1.19. Proposition. *For a shift-bounded measure μ on G and a function $f \in C_c(G)$ the function $\mu * f$ is uniformly continuous.*

Proof. Let $\varepsilon > 0$ be given and choose a compact symmetric neighbourhood V_0 of the neutral element in G. By 1.12 the absolute value $|\mu|$ of μ is shift-bounded and it follows by 1.17 that

$$a = \sup_{x \in G} |\mu|(x + V_0 - \operatorname{supp}(f)) < \infty.$$

The function f being uniformly continuous, there exists a neighbourhood V of the neutral element in G with $V \subseteq V_0$ such that

$$|f(x) - f(y)| \leqq \frac{\varepsilon}{a} \quad \text{for } x, y \in G \text{ with } x - y \in V,$$

and for $x, y \in G$ with $x - y \in V$ we then find

$$\begin{aligned} |\mu * f(x) - \mu * f(y)| &\leqq \int |f(x-z) - f(y-z)| \, d|\mu|(z) \\ &\leqq \frac{\varepsilon}{a} |\mu|(x + V_0 - \operatorname{supp}(f)) \leqq \varepsilon. \quad \square \end{aligned}$$

1.20. Exercise. 1) The mapping $(x, \mu) \mapsto \tau_x \mu$ of $G \times M^+(G)$ into $M^+(G)$ is continuous, when $M^+(G)$ carries the vague topology.

2) Let $f \in C_c(G)$. The mapping $\mu \mapsto \mu * f$ of $M^+(G)$ into $C(G)$ is continuous, when $M^+(G)$ carries the vague topology and $C(G)$ the topology of compact convergence.

1.21. Exercise. 1) The mapping $(x, \mu) \mapsto \tau_x \mu$ of $G \times M_b^+(G)$ into $M_b^+(G)$ is continuous, when $M_b^+(G)$ carries the Bernoulli topology.

2) Let $f \in C_0(G)$. The mapping $\mu \mapsto \mu * f$ of $M_b^+(G)$ into $C_0(G)$ is continuous, when $M_b^+(G)$ carries the Bernoulli topology and $C_0(G)$ the uniform norm topology.

§ 2. Some Basic Results From Harmonic Analysis

The purpose of this paragraph is to give a summary of some of the basic results in harmonic analysis which will be needed in the sequal. For details and proofs we refer to the books of Rudin [1] and Hewitt and Ross [1], [2].

Let G be a LCA-group.

2.1. A *character* on G is a homomorphism γ of G into the *circle group* $\mathbb{T}$ of complex numbers with absolute value 1.

The *dual group* of G is the set $\hat{G}$ of continuous characters on G where the group operation $+$ is defined by

$$(\gamma_1+\gamma_2)(x)=\gamma_1(x)\gamma_2(x) \quad \text{for } \gamma_1, \gamma_2 \in \hat{G} \text{ and } x \in G.$$

The neutral element in $\hat{G}$, denoted 0, is the character: $x \mapsto 1$ and the inverse element of $\gamma \in \hat{G}$ is the character $-\gamma$: $x \mapsto \overline{\gamma(x)}$.

Equipped with the subspace topology from $C(G)$ (having the topology of compact convergence), $\hat{G}$ is a LCA-group.

When K varies over the compact subsets of G and when ε varies in $]0,1[$, the sets

$$U_G(K,\varepsilon)=\{\gamma \in \hat{G} \mid |1-\gamma(x)|<\varepsilon \text{ for all } x \in K\}$$

form a neighbourhood basis for 0 in $\hat{G}$.

In view of Pontryagin's duality theorem, cf. 2.7, stating roughly that G is the dual group of $\hat{G}$, it is customary to write (x,γ) in place of $\gamma(x)$ for $x \in G$ and $\gamma \in \hat{G}$.

When we in the sequel are dealing with a fixed LCA-group G, we will always use the notation Γ for the dual group.

2.2. The maximal ideal space of the commutative Banach algebra $L^1(G)$ is homeomorphic with the dual group Γ, and the Gelfand transformation $\hat{}$ takes the following form

$$\hat{f}(\gamma)=\int \overline{(x,\gamma)}\, f(x)\,dx \quad \text{for } f \in L^1(G) \text{ and } \gamma \in \Gamma.$$

We often write $\mathscr{F}f$ or $\mathscr{F}_G f$ instead of $\hat{f}$, and we call $\mathscr{F}$ ($\mathscr{F}_G$) the *Fourier transformation.*

The Fourier transformation can be extended from $L^1(G)$ to $M_b(G)$, in the literature often called the *Fourier-Stieltjes transformation*, by the definition

$$\hat{\mu}(\gamma)=\int \overline{(x,\gamma)}\, d\mu(x) \quad \text{for } \mu \in M_b(G) \text{ and } \gamma \in \Gamma.$$

As before we also use the notation $\mathscr{F}\mu$ or $\mathscr{F}_G\mu$ instead of $\hat{\mu}$.

It is convenient to introduce the *co-Fourier transformation* $\overline{\mathscr{F}}$ or $\overline{\mathscr{F}}_G$, defined by

$$\overline{\mathscr{F}}\mu(\gamma)=\int (x,\gamma)\,d\mu(x) \quad \text{for } \mu\in M_b(G) \text{ and } \gamma\in\Gamma.$$

Some of the most important properties of $\mathscr{F}$ and $\overline{\mathscr{F}}$ are summed up in the following proposition.

2.3. Proposition. *For $\mu\in M_b(G)$ the Fourier transform $\hat{\mu}$ is a uniformly continuous and bounded function on Γ satisfying*

$$\|\hat{\mu}\|_\infty \leqq \|\mu\|.$$

For $\alpha,\beta\in\mathbb{C}$, $\mu,\nu\in M_b(G)$, $x\in G$ and $\gamma\in\Gamma$ we have $\mathscr{F}\check{\mu}=\overline{\mathscr{F}}\mu$ and $\mathscr{F}\tilde{\mu}=\overline{\mathscr{F}\mu}$, $\mathscr{F}(\alpha\mu+\beta\nu)=\alpha\mathscr{F}\mu+\beta\mathscr{F}\nu$ and $\mathscr{F}(\mu*\nu)=\mathscr{F}\mu\cdot\mathscr{F}\nu$, $\mathscr{F}(\tau_x\mu)(\gamma)=\overline{\gamma(x)}\mathscr{F}\mu(\gamma)$ and $\mathscr{F}(\gamma\mu)=\tau_\gamma\mathscr{F}\mu$.

For $f\in L^1(G)$ the Fourier transform $\hat{f}$ tends to zero at infinity (this property of the Fourier transformation will be referred to as the Riemann-Lebesgue lemma).

The set

$$\{\hat{f}\mid f\in L^1(G)\}$$

is a dense self-adjoint subalgebra of $C_0(\Gamma)$.

The Fourier transformation $\mathscr{F}$ is injective as a mapping from $M_b(G)$.

2.4. Proposition. *For every compact subset $K\subseteq\Gamma$ there exists a function $f\in C_c^+(G)$ such that $\hat{f}\geqq 1_K$.*

Proof. For every $\gamma\in\Gamma$ there exists by Proposition 2.3 a function $h_\gamma\in C_c^+(G)$ such that $\hat{h}_\gamma(\gamma)\neq 0$. The function $g_\gamma=h_\gamma*\tilde{h}_\gamma$ is in $C_c^+(G)$ and it clearly satisfies $\hat{g}_\gamma=|\hat{h}_\gamma|^2\geqq 0$ and $\hat{g}_\gamma(\gamma)>0$. A simple compactness argument now shows the existence of finitely many points $\gamma_1,\dots,\gamma_p$ in Γ such that the function

$$f=\sum_{i=1}^{p} g_{\gamma_i}$$

satisfies $\hat{f}\geqq 0$, $\hat{f}(\gamma)>0$ for all $\gamma\in K$. A positive multiple of f satisfies the condition of the proposition. $\square$

2.5. Theorem (Plancherel). *The Fourier transform $\hat{f}$ of $f\in L^1(G)\cap L^2(G)$ is square integrable with respect to any Haar measure on Γ. There exists a uniquely determined Haar measure ω_Γ on Γ such that*

$$\int |f(x)|^2\,d\omega_G(x)=\int |\hat{f}(\gamma)|^2\,d\omega_\Gamma(\gamma) \quad \textit{for all } f\in L^1(G)\cap L^2(G).$$

The Fourier transformation ^ *extends uniquely by continuity from $L^1(G)\cap L^2(G)$ to $L^2(G)$ as an isometric isomorphism of $L^2(G)$ onto $L^2(\Gamma)$. (G and Γ are equipped with the Haar measure ω_G and ω_Γ.)*

The Haar measure ω_Γ (or just $d\gamma$) on Γ is called the *dual Haar measure* to ω_G on G, and we will always use ω_Γ as the fixed Haar measure on Γ.

2.6. Theorem (The inversion theorem). *Let μ be a bounded measure on G. If the Fourier transform $\hat{\mu}$ belongs to $L^1(\Gamma)$, then the measure μ has a continuous density φ with respect to ω_G given by*

$$\varphi(x)=\int_\Gamma (x,\gamma)\hat{\mu}(\gamma)d\omega_\Gamma(\gamma) \quad \text{for } x\in G.$$

Let G be a LCA-group with Haar measure ω_G, let Γ be the dual group with dual Haar measure ω_Γ and let finally $\hat{\Gamma}$ be the dual group of Γ with dual Haar measure $\omega_{\hat{\Gamma}}$.

We then have a mapping $j_G\colon G\to\hat{\Gamma}$, called the *canonical injection* of G into its bi-dual, defined by

$$(\gamma, j_G(x))=(x,\gamma) \quad \text{for } x\in G \text{ and } \gamma\in\Gamma.$$

2.7. Theorem (Pontryagin's duality theorem). *The canonical injection j_G is an isomorphism and a homeomorphism of G onto $\hat{\Gamma}$ and the image measure $j_G(\omega_G)$ is equal to the Haar measure $\omega_{\hat{\Gamma}}$.*

On account of this theorem we may (and usually do) consider G as the dual group of Γ.

2.8. Definition. For a subset $A\subseteq G$ the *orthogonal complement* of A is the subset $A^\perp\subseteq\Gamma$ given by

$$A^\perp=\{\gamma\in\Gamma \mid (x,\gamma)=1 \text{ for all } x\in A\}.$$

The orthogonal complement $A^\perp$ is a closed subgroup of Γ. Identifying the dual group of Γ with G, the orthogonal complement of a subset $B\subseteq\Gamma$ is the closed subgroup of G, given by

$$B^\perp=\{x\in G \mid (x,\gamma)=1 \text{ for all } \gamma\in B\}.$$

The smallest closed subgroup of G containing a subset A of G is $(A^\perp)^\perp$.

2.9. The mapping $H\mapsto H^\perp$ is a bijection of the set of closed subgroups of G onto the set of closed subgroups of Γ. A subgroup H of G is compact if and only if the orthogonal complement $H^\perp$ is an open subgroup of Γ.

2.10. Let G_1 and G_2 be LCA-groups with dual groups $\hat{G}_1$ and $\hat{G}_2$. For a continuous homomorphism $\varphi\colon G_1\to G_2$ the *dual homomorphism* $\hat{\varphi}\colon \hat{G}_2\to\hat{G}_1$ is defined by

$$(x,\hat{\varphi}(\gamma))=(\varphi(x),\gamma) \quad \text{for } x\in G_1 \text{ and } \gamma\in\hat{G}_2.$$

2.11. Let H be a closed subgroup of G. The *quotient group* G/H is a LCA-group, and the *canonical mapping* $\pi\colon G\to G/H$ of G onto the quotient group G/H is a continuous and open homomorphism. The dual homomorphism $\hat{\pi}\colon \widehat{G/H}\to\hat{G}$ is an isomorphism and a homeomorphism of $\widehat{G/H}$ onto the closed subgroup $H^\perp$ of

$\hat{G}$, and we will therefore consider $H^\perp$ as the dual group of G/H. By duality $\hat{G}/H^\perp$ will be considered as the dual group of H.

2.12. Let G_1 and G_2 be LCA-groups with Haar measures ω_{G_1} and ω_{G_2}. The *product group* $G_1 \times G_2$ is a LCA-group and as Haar measure on $G_1 \times G_2$ we use the product measure $\omega_{G_1} \otimes \omega_{G_2}$.

For $\gamma_1 \in \hat{G}_1$ and $\gamma_2 \in \hat{G}_2$ the mapping

$$(x_1, x_2) \mapsto (x_1, \gamma_1)(x_2, \gamma_2) \tag{1}$$

of $G_1 \times G_2$ into $\mathbb{T}$ is a continuous character on $G_1 \times G_2$. The mapping which to $(\gamma_1, \gamma_2) \in \hat{G}_1 \times \hat{G}_2$ assigns the character (1) is an isomorphism and a homeomorphism of $\hat{G}_1 \times \hat{G}_2$ onto $(G_1 \times G_2)\hat{}$.

§ 3. Positive Definite Functions

In this paragraph we introduce the notion of positive definite functions on a LCA-group G, and we prove the well-known elementary properties of these functions. We state Bochner's theorem but omit the proof which can be found in many textbooks. In Section 13 we prove that the correspondence from Bochner's theorem between positive bounded measures on G and continuous positive definite functions on the dual group Γ is bicontinuous in the appropriate topologies. The classical continuity theorem of Lévy for sequences of positive bounded measures is proved, and we give an example showing that this theorem cannot be extended to nets.

3.1. An $n \times n$ matrix $A = (a_{ij})$ of complex numbers is called *positive hermitian* if

$$\sum_{i,j=1}^{n} a_{ij} c_i \overline{c_j} \geqq 0$$

for all vectors $(c_1, \ldots, c_n) \in \mathbb{C}^n$.

The matrix A is positive hermitian if and only if A is hermitian, i.e. $a_{ij} = \overline{a_{ji}}$, and the eigenvalues of A are non-negative.

3.2. Lemma. *Let $A = (a_{ij})$ and $B = (b_{ij})$ be positive hermitian $n \times n$ matrices. Then the $n \times n$ matrix*

$$C = (c_{ij}) \quad \textit{where} \quad c_{ij} = a_{ij} b_{ij} \quad \textit{for } i, j = 1, \ldots, n,$$

is positive hermitian.

Proof. There exists a positive hermitian matrix $P = (p_{ij})$ such that $B = PP^*$, where P^* is the adjoint matrix of P. We thus have

$$b_{ij} = \sum_{k=1}^{n} p_{ik} \overline{p_{jk}},$$

which implies that for any $(c_1, \ldots, c_n) \in \mathbb{C}^n$

$$\sum_{i,j=1}^{n} a_{ij} b_{ij} c_i \bar{c}_j = \sum_{k=1}^{n} \left(\sum_{i,j=1}^{n} a_{ij} (p_{ik} c_i) \overline{(p_{jk} c_j)} \right) \geqq 0. \quad \square$$

3.3. Definition. A function $\varphi: G \to \mathbb{C}$ is called *positive definite*, if for all $n \in \mathbb{N}$ and for all n-tuples $(x_1, \ldots, x_n)$ of elements from G, the $n \times n$ matrix $(\varphi(x_i - x_j))$ is positive hermitian.

The set of positive definite functions on G is denoted $P(G)$ and the subset of $P(G)$ consisting of the continuous positive definite functions on G is denoted $CP(G)$.

3.4. Let φ be a positive definite function on G.
Since the 1×1 matrix $(\varphi(0))$ is positive hermitian we have $\varphi(0) \geqq 0$.
For every $x \in G$ the matrix

$$\begin{pmatrix} \varphi(0) & \varphi(-x) \\ \varphi(x) & \varphi(0) \end{pmatrix}$$

is positive hermitian, which implies

$$\varphi(-x) = \overline{\varphi(x)}, \quad \text{i.e. } \tilde{\varphi} = \varphi \tag{1}$$

and

$$|\varphi(x)| \leqq \varphi(0). \tag{2}$$

In particular φ is bounded and $\sup_{x \in G} |\varphi(x)| = \varphi(0)$.
For $x, y \in G$ we consider the matrix

$$\begin{pmatrix} \varphi(0) & \overline{\varphi(x)} & \overline{\varphi(y)} \\ \varphi(x) & \varphi(0) & \varphi(x-y) \\ \varphi(y) & \overline{\varphi(x-y)} & \varphi(0) \end{pmatrix}. \tag{3}$$

Supposing that $\varphi(x) \neq \varphi(y)$ we define for $\lambda \in \mathbb{R}$ the triple (c_1, c_2, c_3) by

$$c_1 = 1, \quad c_2 = \frac{\lambda |\varphi(x) - \varphi(y)|}{\varphi(x) - \varphi(y)}, \quad c_3 = -c_2.$$

The positivity of the hermitian matrix (3) then leads to the following inequality

$$\varphi(0)(1 + 2\lambda^2) + 2\lambda |\varphi(x) - \varphi(y)| - 2\lambda^2 \operatorname{Re} \varphi(x-y) \geqq 0,$$

valid for all $\lambda \in \mathbb{R}$, and it follows that the discriminant of the polynomium in λ on the left hand side is $\leqq 0$, i.e.

$$|\varphi(x) - \varphi(y)|^2 \leqq 2\varphi(0)(\varphi(0) - \operatorname{Re} \varphi(x-y)). \tag{4}$$

We conclude (e.g. by (2)), that (4) is valid for all $x, y \in G$.

The determinant of a positive hermitian matrix is non-negative, so if a 3×3 matrix of the form

$$\begin{bmatrix} 1 & \lambda & \mu \\ \bar{\lambda} & 1 & \xi \\ \bar{\mu} & \bar{\xi} & 1 \end{bmatrix}$$

is positive hermitian we find

$$1+\lambda\bar{\mu}\xi+\bar{\lambda}\mu\bar{\xi}\geqq|\lambda|^2+|\mu|^2+|\xi|^2$$

or equivalently

$$|\xi-\bar{\lambda}\mu|^2\leqq(1-|\lambda|^2)(1-|\mu|^2).$$

If this inequality is applied to the matrix (3) for a positive definite function φ satisfying $\varphi(0)=1$, we get

$$|\varphi(x-y)-\varphi(x)\overline{\varphi(y)}|^2\leqq(1-|\varphi(x)|^2)(1-|\varphi(y)|^2).$$

Taking (1) into account we get the following inequality for any positive definite function φ satisfying $\varphi(0)=1$:

$$|\varphi(x+y)-\varphi(x)\varphi(y)|^2\leqq(1-|\varphi(x)|^2)(1-|\varphi(y)|^2). \tag{5}$$

3.5. Any character φ on G is positive definite, because

$$\sum_{i,j=1}^{n}\varphi(x_i-x_j)\,c_i\bar{c}_j=\left|\sum_{i=1}^{n}\varphi(x_i)\,c_i\right|^2\geqq 0.$$

On the other hand any positive definite function $\varphi\colon G\to\mathbb{T}$ is a character on G. This is an immediate consequence of (5).

3.6. Proposition. (i) *The set $P(G)$ is a convex cone which is closed in the topology of pointwise convergence on G.*

(ii) *For $\varphi\in P(G)$ we have $\bar{\varphi}\in P(G)$ and* $\operatorname{Re}\varphi\in P(G)$.

(iii) *For $\varphi_1, \varphi_2\in P(G)$ we have $\varphi_1\varphi_2\in P(G)$.*

(iv) *The non-negative constant functions belong to $P(G)$.*

(v) *The set $CP(G)$ is a convex cone which is a closed subset of $C(G)$ in the topology of compact convergence.*

Proof. Property (iii) follows from Lemma 3.2, and the rest is almost immediate. □

3.7. Exercise. Let $\varphi\colon G\to\mathbb{R}$ satisfy $\varphi=\check{\varphi}$. Then φ is positive definite if (and only if) for all $n\in\mathbb{N}$, for all n-tuples of elements $(x_1,\ldots,x_n)$ from G and for all n-tuples $(c_1,\ldots,c_n)\in\mathbb{Z}^n$ we have

$$\sum_{i,j=1}^{n}\varphi(x_i-x_j)\,c_i c_j\geqq 0.$$

3.8. Exercise. 1) Let H be a non-void subset of G. The characteristic function 1_H of H is positive definite if and only if H is a subgroup of G.

2) Let H be a subgroup of G and let $\varphi \in P(H)$. Then the function

$$\psi(x) = \begin{cases} \varphi(x) & \text{for } x \in H \\ 0 & \text{for } x \notin H, \end{cases}$$

is a positive definite function on G.

3) There is a function $f \in P(\mathbb{R})$, which is Borel measurable but discontinuous at every point of $\mathbb{R}$.

3.9. Exercise. Let $f(z) = \sum_{n=0}^{\infty} a_n z^n$ be a holomorphic function in the disc $|z| < R$ ($R \in]0, \infty]$) and suppose that $a_n \geqq 0$ for all $n \geqq 0$.

For $\varphi \in CP(G)$ with $\varphi(0) < R$ we have $f \circ \varphi \in CP(G)$.

3.10. Proposition. *A positive definite function φ on G is uniformly continuous if* $\operatorname{Re} \varphi$ *is lower semicontinuous at the neutral element.*

Proof. If $\operatorname{Re}\varphi$ is lower semicontinuous at 0, it is in fact continuous at 0, because for $\varepsilon > 0$ we have

$$\{x \in G \mid \operatorname{Re}\varphi(x) \in]\varphi(0)-\varepsilon, \varphi(0)+\varepsilon[\} = \{x \in G \mid \operatorname{Re}\varphi(x) > \varphi(0)-\varepsilon\}$$

on account of the inequality $\operatorname{Re}\varphi(x) \leqq |\varphi(x)| \leqq \varphi(0)$.

From the inequality (4) it follows that φ is uniformly continuous, if $\operatorname{Re}\varphi$ is continuous at the neutral element. □

3.11. Remark. The assumption that G is a LCA-group is not indispensable and the material of Sections 3–10 carries over to the case where G is a (topological) group; then $-x$ and $x-y$ should be replaced by x^{-1} and $y^{-1}x$.

We shall now state the most important result about continuous positive definite functions on a LCA-group G.

3.12. Theorem (Bochner). *A continuous function φ on G is positive definite if and only if there exists a positive bounded measure σ on the dual group Γ such that*

$$\varphi(x) = \int_\Gamma (x, \gamma)\, d\sigma(\gamma) \quad \textit{for } x \in G. \tag{6}$$

The measure σ, which is uniquely determined by (6), *is said to be* associated *with φ.*

For the proof we refer to Rudin [1] or Hewitt and Ross [2].

For $\varphi \in CP(G)$ with associated measure $\sigma \in M_b^+(\Gamma)$ we have

$$\mathscr{F}_\Gamma \check{\sigma}(x) = \varphi(x) \quad \text{for } x \in G,$$

and in particular

$$\sigma(\Gamma) = \varphi(0). \tag{7}$$

Replacing G by Γ, Bochner's theorem can be formulated in the following way:

The Fourier transformation $\mathscr{F}_G$ is a bijection of the cone $M_b^+(G)$ of positive bounded measures on G onto the cone $CP(\Gamma)$ of continuous positive definite functions on Γ.

Taking (7) into account we see that $\mathscr{F}_G$ maps the convex set of probability measures on G onto the convex set of functions $\varphi \in CP(\Gamma)$ verifying $\varphi(0)=1$.

3.13. Theorem. *The Fourier transformation $\mathscr{F}_G$ is a homeomorphism of the cone $M_b^+(G)$ with the Bernoulli topology onto the cone $CP(\Gamma)$ with the topology of compact convergence.*

Proof. Let $(\mu_\alpha)_{\alpha \in A}$ be a net on $M_b^+(G)$ converging in the Bernoulli topology to $\mu \in M_b^+(G)$. Then it is clear that

$$\lim_A \hat{\mu}_\alpha(\gamma) = \hat{\mu}(\gamma) \qquad \text{for all } \gamma \in \Gamma,$$

and we shall prove that the convergence is uniform over compact subsets of Γ.

We start by establishing the following equicontinuity property:

For every $\varepsilon > 0$ there exist a neighbourhood V of the neutral element in Γ and an index $\alpha_0 \in A$ such that for all $\alpha \in A$ and for all $\gamma_1, \gamma_2 \in \Gamma$:

$$\alpha \geqq \alpha_0 \quad \text{and} \quad \gamma_1 - \gamma_2 \in V \quad \text{implies } |\hat{\mu}_\alpha(\gamma_1) - \hat{\mu}_\alpha(\gamma_2)| \leqq \varepsilon. \tag{8}$$

Let $\varepsilon > 0$ be given and choose $\delta > 0$ such that $\delta(3 + \mu(G)) \leqq \varepsilon$, and choose $\varphi \in C_c^+(G)$ satisfying

$$0 \leqq \varphi \leqq 1 \quad \text{and} \quad \int (1-\varphi)\, d\mu < \delta.$$

Since $\lim_A \mu_\alpha = \mu$ in the Bernoulli topology there exists $\alpha_0 \in A$ such that

$$\mu_\alpha(G) < \mu(G) + 1 \quad \text{and} \quad \int (1-\varphi)\, d\mu_\alpha < \delta$$

for all $\alpha \in A$ with $\alpha \geqq \alpha_0$.

Let V be the neighbourhood of 0 in Γ given by, cf. 2.1,

$$V = U_G(\operatorname{supp}(\varphi), \delta).$$

For $\alpha \in A$ and $\gamma_1, \gamma_2 \in \Gamma$ verifying $\alpha \geqq \alpha_0$ and $\gamma_1 - \gamma_2 \in V$ we find

$$\begin{aligned}
|\hat{\mu}_\alpha(\gamma_1) - \hat{\mu}_\alpha(\gamma_2)| &\leqq \int |(x,\gamma_1) - (x,\gamma_2)|\, d\mu_\alpha(x) \\
&\leqq \int |1 - (x, \gamma_1 - \gamma_2)|\, \varphi(x)\, d\mu_\alpha(x) + \int |1 - (x, \gamma_1 - \gamma_2)|\, (1 - \varphi(x))\, d\mu_\alpha(x) \\
&\leqq \delta \int \varphi(x)\, d\mu_\alpha(x) + 2 \int (1 - \varphi(x))\, d\mu_\alpha(x) \\
&\leqq \delta(\mu(G) + 1) + 2\delta \leqq \varepsilon.
\end{aligned}$$

We can now prove that $\hat{\mu}_\alpha$ tends to $\hat{\mu}$ uniformly over compact subsets of Γ.

Let K be a compact subset of Γ and let $\varepsilon>0$ be given. Choose α_0 and V such that (8) holds. Taking limits in "(8)" for $\alpha\in A$ we get

$$|\hat{\mu}(\gamma_1)-\hat{\mu}(\gamma_2)|\leqq\varepsilon$$

for all $\gamma_1, \gamma_2\in\Gamma$ such that $\gamma_1-\gamma_2\in V$.

By compactness there exist finitely many points $\gamma_1, \ldots, \gamma_n\in K$ such that

$$K\subseteq\bigcup_{i=1}^{n}(\gamma_i+V),$$

and then there exist $\alpha_i\in A$, $i=1, \ldots, n$, such that

$$|\hat{\mu}_\alpha(\gamma_i)-\hat{\mu}(\gamma_i)|\leqq\varepsilon \qquad \text{for } \alpha\geqq\alpha_i,\ i=1, \ldots, n.$$

Let $\alpha^*\in A$ be chosen such that $\alpha^*\geqq\alpha_i$ for $i=0, 1, \ldots, n$. For $\gamma\in\gamma_i+V$ and $\alpha\in A$ with $\alpha\geqq\alpha^*$ we then have

$$|\hat{\mu}_\alpha(\gamma)-\hat{\mu}(\gamma)|\leqq|\hat{\mu}_\alpha(\gamma)-\hat{\mu}_\alpha(\gamma_i)|+|\hat{\mu}_\alpha(\gamma_i)-\hat{\mu}(\gamma_i)|+|\hat{\mu}(\gamma_i)-\hat{\mu}(\gamma)|\leqq 3\varepsilon$$

hence

$$\sup_{\gamma\in K}|\hat{\mu}_\alpha(\gamma)-\hat{\mu}(\gamma)|\leqq 3\varepsilon.$$

Let now conversely $(\mu_\alpha)_{\alpha\in A}$ be a net on $M_b^+(G)$, let $\mu\in M_b^+(G)$ and suppose that $\hat{\mu}_\alpha$ tends to $\hat{\mu}$ in the topology of compact convergence. In particular we have

$$\lim_A \mu_\alpha(G)=\lim_A \hat{\mu}_\alpha(0)=\hat{\mu}(0)=\mu(G),$$

so in order to prove that $\lim_A \mu_\alpha=\mu$ in the Bernoulli topology it suffices to verify that $\lim_A \mu_\alpha=\mu$ in the vague topology, cf. Proposition 1.4.

To $\varphi\in C_c^+(G)$ and $\varepsilon>0$ we choose $f\in C_c(\Gamma)$ such that

$$\|\varphi-\mathscr{F}_\Gamma f\|_\infty<\varepsilon,$$

and we then have

$$\begin{aligned}
&|\langle\mu_\alpha, \varphi\rangle-\langle\mu, \varphi\rangle| \\
&\quad\leqq|\langle\mu_\alpha, \varphi-\mathscr{F}_\Gamma f\rangle|+|\langle\mu_\alpha, \mathscr{F}_\Gamma f\rangle-\langle\mu, \mathscr{F}_\Gamma f\rangle|+|\langle\mu, \mathscr{F}_\Gamma f-\varphi\rangle| \\
&\quad\leqq\varepsilon\big(\mu_\alpha(G)+\mu(G)\big)+\int_\Gamma|\hat{\mu}_\alpha(\gamma)-\hat{\mu}(\gamma)|\,|f(\gamma)|\,d\gamma,
\end{aligned}$$

hence

$$\limsup_A|\langle\mu_\alpha, \varphi\rangle-\langle\mu, \varphi\rangle|\leqq 2\varepsilon\,\mu(G),$$

and it follows that $\langle\mu_\alpha, \varphi\rangle$ tends to $\langle\mu, \varphi\rangle$. $\square$

3.14. Theorem (the continuity theorem of Lévy). *Let $(\mu_n)_{n\in\mathbb{N}}$ be a sequence of positive bounded measures on G and let $\varphi: \Gamma \to \mathbb{C}$ be a function on Γ, which is continuous at the neutral element. If*

$$\lim_{n\to\infty} \hat{\mu}_n(\gamma) = \varphi(\gamma) \quad \textit{for all } \gamma \in \Gamma,$$

then there exists a positive bounded measure μ on G such that $(\mu_n)_{n\in\mathbb{N}}$ converges to μ in the Bernoulli topology.

Proof. Since φ is the pointwise limit of positive definite functions, φ is positive definite. By Proposition 3.10 it follows that φ is continuous and by Bochner's theorem there exists $\mu \in M_b^+(G)$ such that $\hat{\mu} = \varphi$.

In analogy with the proof of the preceding theorem it suffices to see that

$$\lim_{n\to\infty} \int |\hat{\mu}_n(\gamma) - \hat{\mu}(\gamma)|\, |f(\gamma)|\, d\gamma = 0$$

for all $f \in C_c(\Gamma)$, and this is a consequence of Lebesgue's theorem on dominated convergence. □

The continuity theorem of Lévy can not be extended to nets, as the following example shows.

3.15. Example. Let K be the compact group $\mathbb{T}^{\mathbb{R}}$ and let j: $\mathbb{R} \to K$ be defined by

$$j(x) = (e^{-itx})_{t\in\mathbb{R}}.$$

Then j is a continuous injective homomorphism, but j is not an isomorphism of $\mathbb{R}$ onto $j(\mathbb{R})$. In fact, if j was an isomorphism, then $j(\mathbb{R})$ would be a locally compact subgroup of the compact group K, and then $j(\mathbb{R})$ would be compact, cf. Hewitt and Ross [1], p. 35.

This proves the existence of a net $(y_\alpha)_{\alpha\in A}$ in $j(\mathbb{R})$ such that $\lim_A y_\alpha = j(0)$, but such that the net $(x_\alpha)_{\alpha\in A}$ on $\mathbb{R}$ defined by $x_\alpha = j^{-1}(y_\alpha)$ does not converge to 0 in $\mathbb{R}$.

The corresponding net $(\varepsilon_{x_\alpha})_{\alpha\in A}$ on $M_b^+(\mathbb{R})$ does not converge to ε_0 in the Bernoulli topology but the Fourier transforms (cf. § 5)

$$\hat{\varepsilon}_{x_\alpha}: t \mapsto e^{-itx_\alpha}$$

converge pointwise to the function $\hat{\varepsilon}_0: t \mapsto 1$.

§ 4. Fourier Transformation of Positive Definite Measures

The main topic of this paragraph is the study of positive definite measures on the LCA-group G. After the definition of positive definite measures, the Fourier transform of a positive definite measure is introduced as a positive measure on the dual group Γ. The existence of this Fourier transform is established by

Bochner's theorem. The theory of Fourier transformation of positive definite measures allows a very natural approach to the Plancherel theorem and other parts of harmonic analysis, cf. Remark. 4.13.

4.1. Proposition. *A continuous function $\varphi\colon G\to\mathbb{C}$ is positive definite if and only if*

$$\int \varphi(x)\, f*\tilde{f}(x)\,dx \geqq 0 \quad \textit{for all } f\in C_c(G).$$

Proof. For an n-tuple $(x_1,\dots,x_n)$ of elements from G and an n-tuple $(c_1,\dots,c_n)\in\mathbb{C}^n$ we define the discrete measure

$$\mu=\sum_{i=1}^{n} c_i\varepsilon_{x_i}, \tag{1}$$

and then we have

$$\sum_{i,j=1}^{n} \varphi(x_i-x_j)\,c_i\bar{c}_j=\int \varphi\, d(\mu*\tilde{\mu}).$$

The proposition now follows, because for every $f\in C_c(G)$ the measure $f\omega_G$ is vague limit of measures of the form (1) with $c_i\in\mathbb{C}$ and $x_i\in\operatorname{supp}(f)$, and because every measure μ of the form (1) is vague limit of measures $f\omega_G$, where $f\in C_c(G)$ has its support in a fixed compact neighbourhood of the set $\{x_1,\dots,x_n\}$. □

4.2. Definition. A measure μ on G is called *positive definite* if

$$\langle \mu, f*\tilde{f}\rangle \geqq 0 \quad \text{for all } f\in C_c(G).$$

The set of positive definite measures on G is denoted $M_p(G)$; it is clear that $M_p(G)$ is a vaguely closed convex cone in $M(G)$, which is stable under reflection and conjugation, i.e. $\mu\in M_p(G)$ implies $\check{\mu}\in M_p(G)$ and $\bar{\mu}\in M_p(G)$.

By Proposition 4.1 a continuous function is density for a positive definite measure if and only if it is positive definite.

If μ is a measure on G such that the convolution of μ and $\tilde{\mu}$ exists, then the measure $\mu*\tilde{\mu}$ is positive definite.

The Dirac measure ε_0 and the Haar measure ω_G are positive definite, because for every $f\in C_c(G)$

$$\langle \varepsilon_0, f*\tilde{f}\rangle=\int |f(x)|^2\,dx \quad\text{and}\quad \langle \omega_G, f*\tilde{f}\rangle=|\langle \omega_G, f\rangle|^2.$$

4.3. Proposition. *Let φ be a continuous positive definite function on G and let $\sigma\in M_b^+(\Gamma)$ be the associated measure.*

For any $f\in C_c(G)$ the function $\varphi*f*\tilde{f}$ is continuous and positive definite and the associated measure is $|\hat{f}|^2\sigma$.

Proof. It is enough to show that the co-Fourier transform of the positive bounded measure $|\hat{f}|^2 \sigma$ is $\varphi * f * \tilde{f}$:

$$\begin{aligned}\int (x,\gamma)|\hat{f}(\gamma)|^2 d\sigma(\gamma) &= \int \Big((x,\gamma) \int \overline{(y,\gamma)}\, f * \tilde{f}(y)\, dy\Big) d\sigma(\gamma)\\ &= \int \Big(\int (y,\gamma)\, f * \tilde{f}(x-y)\, dy\Big) d\sigma(\gamma) = \int \varphi(y)\, f * \tilde{f}(x-y)\, dy = \varphi * f * \tilde{f}(x). \quad \square\end{aligned}$$

4.4. Proposition. *A measure μ on G is positive definite if and only if $\mu * f * \tilde{f}$ is a continuous positive definite function for all $f \in C_c(G)$.*

A positive and positive definite measure μ on G is shift-bounded.

Proof. Suppose first that μ is positive definite and let $f, g \in C_c(G)$. The function $\mu * f * \tilde{f}$ is clearly continuous and we have

$$\int \mu * f * \tilde{f}(x)\, g * \tilde{g}(x)\, dx = \langle \mu, (\check{f} * g) * (\check{f} * g)^{\sim} \rangle \geqq 0.$$

By Proposition 4.1 it follows that $\mu * f * \tilde{f} \in CP(G)$.

On the other hand, if $\mu * f * \tilde{f} \in CP(G)$ for $f \in C_c(G)$, then in particular $\mu * f * \tilde{f}(0) \geqq 0$, i.e.

$$\langle \mu, \check{f} * \tilde{\check{f}} \rangle \geqq 0,$$

which shows that μ is positive definite.

Let $\mu \in M_p^+(G)$ and $f \in C_c^+(G)$. There exists $g \in C_c^+(G)$ such that $f \leqq g * \tilde{g}$ and hence

$$\mu * f \leqq \mu * g * \tilde{g}.$$

The function $\mu * g * \tilde{g}$ is positive definite and therefore bounded, and this implies that $\mu * f$ is bounded. $\square$

4.5. Theorem. *Let μ be a positive definite measure on G.*

There exists one and only one measure σ on Γ such that

(i) $\int |\hat{f}|^2 d|\sigma| < \infty$ *for* $f \in C_c(G)$,

(ii) $\mu * f * \tilde{f}(x) = \int (x,\gamma) |\hat{f}(\gamma)|^2 d\sigma(\gamma)$ *for* $f \in C_c(G)$ *and* $x \in G$.

The uniquely determined measure σ on Γ verifying (i) *and* (ii) *is positive, and it is called the* measure associated with μ.

Proof. It follows by Proposition 4.4 that $\mu * f * \tilde{f} \in CP(G)$ for every $f \in C_c(G)$. The positive bounded measure on Γ associated with $\mu * f * \tilde{f}$ is denoted σ_f, and by Proposition 4.3 we then have for all $f, g \in C_c(G)$

$$|\hat{f}|^2 \sigma_g = |\hat{g}|^2 \sigma_f, \tag{2}$$

because both measures are associated with the function $\mu * f * \tilde{f} * g * \tilde{g} \in CP(G)$, and hence identical.

Suppose that σ is a measure on Γ satisfying (i) and (ii). We then have

$$|\hat{f}|^2\sigma=\sigma_f \tag{3}$$

for all $f\in C_c(G)$. By (3) we see that $\langle\sigma,\varphi\rangle$ is uniquely determined for every $\varphi\in C_c(\Gamma)$. We only have to choose $f\in C_c(G)$ such that $\hat{f}(\gamma)\neq 0$ for all $\gamma\in\operatorname{supp}(\varphi)$ (this is possible, cf. 2.4), and then we have

$$\langle\sigma,\varphi\rangle=\int\frac{\varphi}{|\hat{f}|^2}\,d\sigma_f,$$

in particular that σ is a positive measure.

In the above formula $\dfrac{\varphi}{|\hat{f}|^2}$ stands for the function

$$\gamma\mapsto\begin{cases}\dfrac{\varphi(\gamma)}{|\hat{f}(\gamma)|^2} & \text{if } \hat{f}(\gamma)\neq 0,\\ 0 & \text{if } \hat{f}(\gamma)=0,\end{cases}$$

which is continuous with compact support.

In order to define $\langle\sigma,\varphi\rangle$ for $\varphi\in C_c(\Gamma)$ we choose $f\in C_c(G)$ such that $\hat{f}(\gamma)\neq 0$ for all $\gamma\in\operatorname{supp}(\varphi)$, and then we put

$$\langle\sigma,\varphi\rangle=\int\frac{\varphi}{|\hat{f}|^2}\,d\sigma_f. \tag{4}$$

If $g\in C_c(G)$ is another function with the property that $\hat{g}(\gamma)\neq 0$ for all $\gamma\in\operatorname{supp}(\varphi)$ we have by (2)

$$\int\frac{\varphi}{|\hat{f}|^2}\,d\sigma_f=\int\frac{\varphi}{|\hat{g}|^2}\,d\sigma_g,$$

so the number $\langle\sigma,\varphi\rangle$ defined by (4) is independent of f. It is easy to see that σ is a positive linear form on $C_c(\Gamma)$, i.e. a positive measure on Γ.

It remains to see that σ verifies the conditions (i) and (ii), but this follows as soon as we have proved that

$$|\hat{f}|^2\sigma=\sigma_f \quad \text{for } f\in C_c(G).$$

Let $\varphi\in C_c(\Gamma)$ be given and choose $g\in C_c(G)$ such that $\hat{g}(\gamma)\neq 0$ for all $\gamma\in\operatorname{supp}(\varphi)$. By (4) and (2) we then have

$$\begin{aligned}\langle|\hat{f}|^2\sigma,\varphi\rangle&=\langle\sigma,|\hat{f}|^2\varphi\rangle\\&=\int\frac{|\hat{f}|^2\varphi}{|\hat{g}|^2}\,d\sigma_g=\int\frac{\varphi|\hat{g}|^2}{|\hat{g}|^2}\,d\sigma_f=\langle\sigma_f,\varphi\rangle.\quad\square\end{aligned}$$

4.6. Remark. It follows by Proposition 4.3 that the notion of a measure associated with a positive definite measure is an extension of the previous notion (cf. 3.12) of a measure associated with a continuous positive definite function.

4.7. Theorem. *A measure μ on G is positive definite if and only if there exists a positive measure σ on Γ such that*

$$\langle\mu, f*\tilde{f}\rangle = \int |\bar{\mathscr{F}}_G f|^2\, d\sigma \tag{5}$$

for $f\in C_c(G)$, and σ is in the affirmative case the measure associated with μ.

Proof. Let μ be a measure on G for which there exists $\sigma\in M^+(\Gamma)$ such that (5) holds for all $f\in C_c(G)$. It is clear that μ is positive definite. Let $f\in C_c(G)$ and $x\in G$ be given. Replacing f by $\check{f}$ in (5) we get

$$\int |\bar{\mathscr{F}}_G f|^2\, d\sigma = \langle\mu, \check{f}*\bar{f}\rangle < \infty,$$

i.e. condition (i) of Theorem 4.5 is satisfied. Furthermore by polarization of (5) we find

$$\langle\mu, f*\tilde{g}\rangle = \int \bar{\mathscr{F}}_G f\, \overline{\bar{\mathscr{F}}_G g}\, d\sigma \qquad \text{for } g\in C_c(G).$$

Replacing f by $\tau_x \bar{f}$ and g by $\bar{f}$, we see that condition (ii) of Theorem 4.5 is verified.

If μ is positive definite, then the associated measure σ satisfies (5) for $f\in C_c(G)$. In fact, replacing f by $\check{f}$ and putting $x=0$ in condition (ii) of Theorem 4.5 we get (5). □

4.8. Corollary. *The mapping taking a positive definite measure into its associated measure is injective.*

Proof. Let $\mu, \nu\in M_p(G)$ and suppose that $\sigma\in M^+(\Gamma)$ is associated measure for μ and ν. By Theorem 4.7 we then have

$$\langle\mu, f*\tilde{f}\rangle = \langle\nu, f*\tilde{f}\rangle \qquad \text{for } f\in C_c(G).$$

By polarization of this equation we get

$$\langle\mu, f*\tilde{g}\rangle = \langle\nu, f*\tilde{g}\rangle \qquad \text{for } f, g\in C_c(G),$$

and letting g run through an approximate unit we find, cf. 1.18, that

$$\langle\mu, f\rangle = \langle\nu, f\rangle \qquad \text{for } f\in C_c(G). \quad \square$$

4.9. Proposition. *For a positive definite measure μ on G the associated measure σ on Γ is shift-bounded. If furthermore μ is absolutely continuous with respect to the Haar measure on G, then σ tends to zero at infinity.*

Proof. For every $\gamma \in \Gamma$ and all $f \in C_c(G)$ we have

$$\gamma[f * \tilde{f}] = [\gamma f] * [\gamma f]^{\sim}. \tag{6}$$

Replacing f by $\bar{\gamma} f$ in (5) we get on account of (6)

$$\int \overline{(x, \gamma)} f * \tilde{f}(x) d\mu(x) = \int |\hat{f}(\gamma - \delta)|^2 d\sigma(\delta) = \sigma * |\hat{f}|^2(\gamma).$$

The measure $(f * \tilde{f})\mu$, having compact support, is bounded, and its Fourier transform $\sigma * |\hat{f}|^2$ is consequently bounded. Since every function $\varphi \in C_c^+(\Gamma)$ can be majorized by some function $|\hat{f}|^2$ with $f \in C_c(G)$ (cf. 2.4), we have that $\sigma * \varphi$ is bounded for all $\varphi \in C_c^+(\Gamma)$.

If μ is absolutely continuous with respect to the Haar measure on G, then the measure $(f * \tilde{f})\mu$ has an integrable density, and its Fourier transform $\sigma * |\hat{f}|^2$ tends therefore to zero at infinity. By an argument as above, this ensures that σ tends to zero at infinity. □

4.10. Proposition. *Let μ be a positive definite measure on G with associated measure σ on Γ and let φ be a continuous positive definite function on G with associated measure τ on Γ. The measure $\varphi\mu$ is positive definite and the convolution of the measures σ and τ exists and $\sigma * \tau$ is the measure associated with $\varphi\mu$.*

Proof. The convolution $\sigma * \tau$ exists because σ is shift-bounded and τ is bounded, cf. 1.13.

For $f \in C_c(G)$ we get by (6) and (5)

$$\begin{aligned}\int \varphi(x) f * \tilde{f}(x) d\mu(x) &= \int (f * \tilde{f}(x) \int (x, \gamma) d\tau(\gamma)) d\mu(x)\\ &= \int \langle \mu, (\gamma f) * (\gamma f)^{\sim} \rangle d\tau(\gamma)\\ &= \int (\int |\mathscr{F}_G(\gamma f)|^2 d\sigma) d\tau(\gamma)\\ &= \int\int |\mathscr{F}_G f|^2 (\gamma + \eta) d\sigma(\eta) d\tau(\gamma).\end{aligned}$$

Since the last expression is non-negative, $\varphi\mu$ is a positive definite measure. It follows from Theorem 4.7 that the measure associated with $\varphi\mu$ is $\sigma * \tau$. □

4.11. Corollary. *Let μ be a positive definite measure on G with associated measure σ. The periodicity group for σ (cf. 6.1), i.e. the set*

$$\operatorname{per}(\sigma) = \{\gamma \in \Gamma \mid \tau_\gamma \sigma = \sigma\}$$

is the orthogonal complement of the support of μ, i.e.

$$\operatorname{per}(\sigma) = \{\gamma \in \Gamma \mid \gamma = 1 \text{ on } \operatorname{supp}(\mu)\}.$$

Proof. The measure associated with the continuous positive definite function $\gamma \in \Gamma$ is ε_γ, and the measure associated with $\gamma\mu$ is therefore $\varepsilon_\gamma * \sigma = \tau_\gamma \sigma$. It follows by Corollary 4.8 that $\gamma \in \operatorname{per}(\sigma)$ if and only if $\gamma\mu = \mu$, and this is equivalent with γ being 1 on the support of μ. □

4.12. Remark. The Dirac measure ε_0 is positive definite and the associated measure σ has by Corollary 4.11 all characters $\gamma\in\Gamma$ as periods, i.e. σ is a Haar measure on Γ. By Theorem 4.5 we have

$$f*\tilde{f}(x)=\int (x,\gamma)|\hat{f}(\gamma)|^2 d\sigma(\gamma) \tag{7}$$

for all $f\in C_c(G)$ and $x\in G$, in particular for $x=0$

$$\int |f(x)|^2 dx=\int |\hat{f}(\gamma)|^2 d\sigma(\gamma). \tag{8}$$

It follows that σ is the Haar measure on Γ dual to ω_G, i.e. $\sigma=\omega_\Gamma$, cf. 2.5.

4.13. Remark. Although we have assumed Plancherel's theorem, the inversion theorem and Pontryagin's theorem to be known in this exposition we want to point out that harmonic analysis on LCA-groups can be built up in the following way: One follows the exposition of Rudin [1] as far as the theorem of Bochner. By means of this theorem the theory of Fourier transformation of positive definite measures is developed. Plancherel's theorem and the inversion theorem are easy corollaries of this theory. Finally one deduces Pontryagin's theorem in the usual way from the Plancherel theorem.

4.14. Proposition. *A bounded measure μ on G is positive definite if and only if $\hat{\mu}(\gamma)\geqq 0$ for all $\gamma\in\Gamma$.*

If $\mu\in M_b(G)$ is positive definite, the associated measure is $\hat{\mu}\omega_\Gamma$, where ω_Γ is the Haar measure on Γ dual to ω_G.

Proof. Since $\mu\in M_b(G)$ we have

$$|\hat{\mu}(\gamma)|\leqq\|\mu\| \quad \text{for all } \gamma\in\Gamma,$$

and therefore

$$\int |\hat{f}(\gamma)|^2|\hat{\mu}(\gamma)|\,d\omega_\Gamma(\gamma)\leqq\|\mu\|\int |\hat{f}(\gamma)|^2 d\omega_\Gamma(\gamma)<\infty \quad \text{for } f\in C_c(G).$$

It follows by (7) (with σ replaced by ω_Γ) that for all $f\in C_c(G)$ and $x\in G$

$$\mu*f*\tilde{f}(x)=\int (x,\gamma)|\hat{f}(\gamma)|^2\hat{\mu}(\gamma)\,d\omega_\Gamma(\gamma). \tag{9}$$

This shows that the measure $\hat{\mu}\omega_\Gamma$ fulfils the conditions of Theorem 4.5, so if μ is positive definite, then $\hat{\mu}\omega_\Gamma$ is the associated measure and hence positive. This implies that the continuous function $\hat{\mu}$ is non-negative.

If conversely $\hat{\mu}$ is non-negative it follows by (9) that

$$\langle\mu,f*\tilde{f}\rangle=\int |\mathscr{F}_G f(\gamma)|^2\hat{\mu}(\gamma)\,d\omega_\Gamma(\gamma)\geqq 0 \quad \text{for } f\in C_c(G),$$

and μ is consequently positive definite. □

Proposition 4.14 justifies the following definition.

4.15. Definition. The *Fourier transform* $\mathscr{F}\mu$ or $\mathscr{F}_G\mu$ of a positive definite measure μ on G is the measure on Γ associated with μ.

We can consider the Fourier transformation as a mapping from $M_p(G)$ into $M^+(\Gamma)$ and this mapping clearly satisfies

$$\mathscr{F}_G(\alpha\mu+\beta\nu)=\alpha\mathscr{F}_G\mu+\beta\mathscr{F}_G\nu$$

for $\alpha, \beta \geqq 0$ and $\mu, \nu \in M_p(G)$. Furthermore $\mathscr{F}_G$ is injective by Corollary 4.8.

4.16. Theorem. *The Fourier transformation $\mathscr{F}_G$ is a bijective mapping of the cone $M_p^+(G)$ of positive and positive definite measures on G onto the cone $M_p^+(\Gamma)$, and the inverse mapping is $\mathscr{F}_\Gamma$. Furthermore $\mathscr{F}_G$ is a homeomorphism, when $M_p^+(G)$ and $M_p^+(\Gamma)$ are carrying the vague topologies.*

Proof. Let $\varphi \in CP^+(G)$ and let $\sigma=\mathscr{F}_G\varphi$ be the associated measure. We then have

$$\mathscr{F}_\Gamma\sigma(x)=\overline{\mathscr{F}}_\Gamma\sigma(-x)=\varphi(-x)=\varphi(x) \quad \text{for } x\in G,$$

and it follows by Proposition 4.14 that σ is positive definite.

This proves that $\mathscr{F}_G$ maps $CP^+(G)$ into $M_p^+(\Gamma)$ and that $\mathscr{F}_\Gamma\mathscr{F}_G$ is the identity mapping on $CP^+(G)$.

Let $\mu\in M_p^+(G)$. We shall prove that $\mathscr{F}_G\mu\in M_p(\Gamma)$. For an approximate unit $(\varphi_V)_{V\in\check{V}}$ we have

$$\lim_{\check{V}} \hat{\varphi}_V=1 \tag{10}$$

uniformly over compact subsets of Γ. This can be seen directly, but follows also from 3.13 since $\varphi_V\,\omega_G$ tends to ε_0 in the Bernoulli topology. For every $V\in\check{V}$ the function

$$\mu_V=\mu*\varphi_V*\tilde{\varphi}_V$$

belongs to $CP^+(G)$, and its Fourier transform

$$\mathscr{F}_G\mu_V=|\hat{\varphi}_V|^2\mathscr{F}_G\mu$$

belongs therefore to $M_p^+(\Gamma)$. We have consequently

$$\int g*\tilde{g}\,|\hat{\varphi}_V|^2\,d\mathscr{F}_G\mu\geqq 0 \tag{11}$$

for all $g\in C_c(\Gamma)$ and all $V\in\check{V}$. Going to the limit in (11) for V running through $\check{V}$, we get by (10)

$$\int g*\tilde{g}\,d\mathscr{F}_G\mu\geqq 0 \quad \text{for all } g\in C_c(\Gamma),$$

hence $\mathscr{F}_G\mu\in M_p(\Gamma)$.

By Proposition 4.10 with G replaced by Γ we have

$$\mathscr{F}_\Gamma(\mathscr{F}_G \mu_V) = \mathscr{F}_\Gamma(|\hat{\varphi}_V|^2 \mathscr{F}_G \mu) = \varphi_V * \tilde{\varphi}_V * \mathscr{F}_\Gamma \mathscr{F}_G \mu,$$

and on the other hand

$$\mathscr{F}_\Gamma(\mathscr{F}_G \mu_V) = \mu_V = \varphi_V * \tilde{\varphi}_V * \mu,$$

because $\mu_V \in CP^+(G)$. It follows that

$$\varphi_V * \tilde{\varphi}_V * \mathscr{F}_\Gamma \mathscr{F}_G \mu = \varphi_V * \tilde{\varphi}_V * \mu,$$

and taking limits of both sides for $V \in \check{V}$, we find

$$\mu = \mathscr{F}_\Gamma \mathscr{F}_G \mu.$$

It remains to see that $\mathscr{F}_G$ is a homeomorphism.

Let $(\mu_\alpha)_{\alpha \in A}$ be a net on $M_p^+(G)$ converging vaguely to $\mu \in M_p^+(G)$. By Exercise 1.20 we have

$$\lim_A \mu_\alpha * g = \mu * g \quad \text{in } C(G) \text{ for all } g \in C_c(G),$$

and in particular

$$\lim_A \mu_\alpha * f * \tilde{f} = \mu * f * \tilde{f} \quad \text{in } C(G) \text{ for all } f \in C_c(G).$$

By Theorem 3.13 we get for every $f \in C_c(G)$ that

$$\lim_A |\hat{f}|^2 \mathscr{F}_G \mu_\alpha = \lim_A \mathscr{F}_G(\mu_\alpha * f * \tilde{f}) = \mathscr{F}_G(\mu * f * \tilde{f}) = |\hat{f}|^2 \mathscr{F}_G \mu$$

in the Bernoulli topology.

For given $\varphi \in C_c(\Gamma)$ we choose $f \in C_c(G)$ such that $\hat{f}(\gamma) \neq 0$ for all $\gamma \in \operatorname{supp}(\varphi)$. The function

$$h(\gamma) = \begin{cases} \dfrac{\varphi(\gamma)}{|\hat{f}(\gamma)|^2} & \text{if } \hat{f}(\gamma) \neq 0, \\ 0 & \text{if } \hat{f}(\gamma) = 0, \end{cases}$$

belongs to $C_c(\Gamma)$, and we therefore have

$$\lim_A \langle |\hat{f}|^2 \mathscr{F}_G \mu_\alpha, h \rangle = \langle |\hat{f}|^2 \mathscr{F}_G \mu, h \rangle$$

or

$$\lim_A \langle \mathscr{F}_G \mu_\alpha, \varphi \rangle = \langle \mathscr{F}_G \mu, \varphi \rangle,$$

which proves that $\mathscr{F}_G \mu_\alpha$ tends to $\mathscr{F}_G \mu$ vaguely.

So far we have proved that $\mathscr{F}_G$ is a continuous mapping of $M_p^+(G)$ into $M_p^+(\Gamma)$ such that $\mathscr{F}_\Gamma \mathscr{F}_G$ is the identity on $M_p^+(G)$. Interchanging the roles of G and Γ we obtain the statement of the theorem. □

4.17. Exercise. Every $\mu \in M_p(G)$ has the property $\mu = \tilde{\mu}$ and every $\mu \in M_p^+(G)$ is symmetric. For $\mu \in M_p(G)$ we have $\mathscr{F}_G \check{\mu} = (\mathscr{F}_G \mu)\check{}$.

4.18. Exercise. 1) Let $\mu \in M_p(G)$ and $\nu \in M_p(G) \cap M_c(G)$. Then

$$\mu * \nu \in M_p(G) \quad \text{and} \quad \mathscr{F}_G(\mu * \nu) = \hat{\nu} \mathscr{F}_G \mu.$$

2) Let $\mu \in M_p^+(G)$ and $\nu \in M_p(G) \cap M_b(G)$. Then the convolution $\mu * \nu$ exists, $\mu * \nu \in M_p(G)$ and $\mathscr{F}_G(\mu * \nu) = \hat{\nu} \mathscr{F}_G \mu$.

4.19. The Fourier transform of a positive definite measure is defined in Godement [1] by means of a general "Plancherel theorem" which includes the classical Plancherel theorem and Bochner's theorem.

The present method to define the Fourier transform of a positive definite measure is introduced in Berg [1].

§ 5. Positive Definite Functions on $\mathbb{R}$

In this paragraph we present some classical examples of positive definite functions on the real line together with their associated measures. Furthermore we give a proof of a theorem due to Polya, stating that an even continuous function $\varphi: \mathbb{R} \to [0, \infty[$, which is decreasing and convex on $]0, \infty[$, is positive definite.

5.1. The additive group of real numbers with the usual topology is a LCA-group. For every $y \in \mathbb{R}$ the function

$$x \longmapsto e^{ixy}$$

is a continuous character on $\mathbb{R}$ and the mapping

$$y \longmapsto (x \longmapsto e^{ixy}) \tag{1}$$

is an isomorphism and a homeomorphism of $\mathbb{R}$ onto the dual group $\hat{\mathbb{R}}$. We will always consider $\mathbb{R}$ as the dual group of $\mathbb{R}$ because of (1).

As Haar measure on $\mathbb{R}$ we use the Lebesgue measure $\omega_{\mathbb{R}}$. The dual Haar measure on $\mathbb{R}$ (considered as the dual group of $\mathbb{R}$) is $\frac{1}{2\pi}\omega_{\mathbb{R}}$.

The Fourier transform of a measure $\mu \in M_b(\mathbb{R})$ is the function $\hat{\mu}$ on $\mathbb{R}$ defined by

$$\hat{\mu}(y) = \int_{-\infty}^{\infty} e^{-ixy} d\mu(x) \quad \text{for } y \in \mathbb{R}. \tag{2}$$

If μ is a positive bounded measure on ℝ, then $\hat{\mu}$ is a continuous positive definite function on ℝ, and every such function is, by Bochner's theorem, the Fourier transform of a positive bounded measure on ℝ.

5.2. The following table contains some important probability measures μ on ℝ and their Fourier transforms $\hat{\mu}$, which are continuous positive definite functions on ℝ taking the value 1 at the origin. The independent variable in $\hat{\mu}$ is denoted y.

Table 1

Probability measure	μ	$\hat{\mu}$
Degenerate distribution	ε_a	e^{-iay}
Symmetric degenerate distribution	$\frac{1}{2}(\varepsilon_a+\varepsilon_{-a})$	$\cos(ay)$
Binomial distribution with parameter $p\in]0,1[$, $q=1-p$	$\sum_{j=0}^{n}\binom{n}{j}p^j q^{n-j}\varepsilon_j$	$(q+pe^{-iy})^n$
Poisson distribution with parameter $\lambda>0$	$\sum_{n=0}^{\infty}e^{-\lambda}\frac{\lambda^n}{n!}\varepsilon_n$	$\exp(\lambda(e^{-iy}-1))$
Uniform distribution on the interval $[-a,a]$, $a>0$	$\frac{1}{2a}1_{[-a,a]}(x)\,dx$	$\frac{\sin(ay)}{ay}$
Triangular distribution on the interval $[-a,a]$, $a>0$	$\frac{1}{a}\left(1-\frac{\lvert x\rvert}{a}\right)1_{[-a,a]}(x)\,dx$	$\frac{2(1-\cos(ay))}{a^2y^2}$
Laplace distribution with parameter $\sigma>0$	$\frac{1}{2\sigma}\exp\left(-\frac{\lvert x\rvert}{\sigma}\right)dx$	$(1+\sigma^2y^2)^{-1}$
Normal distribution with parameter $t>0$	$\frac{1}{\sqrt{4\pi t}}\exp\left(-\frac{x^2}{4t}\right)dx$	$\exp(-ty^2)$
Cauchy distribution with parameter $t>0$	$\frac{t}{\pi}(t^2+x^2)^{-1}\,dx$	$e^{-t\lvert y\rvert}$
Gamma distribution with parameter $t>0$	$\frac{1}{\Gamma(t)}x^{t-1}e^{-x}1_{]0,\infty[}(x)\,dx$	$(1+iy)^{-t}$

The next proposition shows that $\varphi(x)$ cannot tend too fast to $\varphi(0)$ when $x\to 0$ for a non-constant function $\varphi\in CP(\mathbb{R})$.

5.3. Proposition. *A continuous positive definite function φ on ℝ verifying*

$$\lim_{x\to 0}\frac{\varphi(0)-\varphi(x)}{x^2}=0.$$

is constant.

Proof. By Bochner's theorem there exists a positive bounded measure σ on $\mathbb{R}$ such that

$$\varphi(x)=\int e^{ixy}\,d\sigma(y) \quad \text{for } x\in\mathbb{R}.$$

For $x\neq 0$ we then have

$$\frac{2\varphi(0)-\varphi(x)-\varphi(-x)}{x^2}=\int\frac{2-e^{ixy}-e^{-ixy}}{x^2}\,d\sigma(y)$$
$$=\int\left(\frac{\sin\frac{xy}{2}}{\frac{xy}{2}}\right)^2 y^2\,d\sigma(y).$$

By Fatou's lemma we find

$$0\leqq\int y^2\,d\sigma(y)=\int\lim_{x\to 0}\left(\frac{\sin\frac{xy}{2}}{\frac{xy}{2}}\right)^2 y^2\,d\sigma(y)$$
$$\leqq\liminf_{x\to 0}\int\left(\frac{\sin\frac{xy}{2}}{\frac{xy}{2}}\right)^2 y^2\,d\sigma(y)=\liminf_{x\to 0}\frac{2\varphi(0)-\varphi(x)-\varphi(-x)}{x^2}=0$$

by assumption. It follows that

$$\int y^2\,d\sigma(y)=0,$$

which implies that $\operatorname{supp}(\sigma)\subseteq\{0\}$, hence $\sigma=\varphi(0)\varepsilon_0$ and φ is constant. □

We shall now prove an important criterion, involving convexity, for positive definiteness. As a preparation we define a family $(\varphi_a)_{a>0}$ of functions $\varphi_a\colon\mathbb{R}\to\mathbb{R}$ given by

$$\varphi_a(x)=\begin{cases}1-\dfrac{|x|}{a} & \text{for } |x|\leqq a,\\[2mm] 0 & \text{for } |x|>a.\end{cases}$$

Since the Fourier transform $\hat{\varphi}_a$ is a positive function (cf. the above table), Propositions 4.1 and 4.14 show that φ_a is positive definite.

5.4. Theorem (Polya). *A continuous function $\varphi\colon\mathbb{R}\to[0,\infty[$ verifying*

(i) *φ is even, i.e. $\varphi(-x)=\varphi(x)$,*

(ii) *φ is decreasing and convex on the interval $]0,\infty[$,*

is positive definite.

Proof. The limit $\alpha = \lim_{x\to\infty} \varphi(x)$ exists and $\alpha \geqq 0$. Without loss of generality we may assume that $\varphi(0)>0$ and $\alpha=0$. Then φ is strictly decreasing. We shall see that φ can be approximated uniformly with sums of positive multiples of the functions φ_a above. Let $\varepsilon \in]0, \varphi(0)[$ be given. There exists $x_0>0$ such that $\varphi(x_0)=\varepsilon$. For $x\in[0,x_0[$ we have $\varphi(x)>\varepsilon$ and for $x\in]x_0,\infty[$ we have $\varphi(x)<\varepsilon$. Since φ is uniformly continuous on the interval $[0,x_0]$, there exist numbers a_i with $0=a_0<a_1<\dots<a_{n-1}=x_0$ such that

$$|\varphi(x)-\varphi(y)|<\varepsilon \quad \text{for } x,y\in[a_k,a_{k+1}],\ k=0,1,\dots,n-2.$$

Putting $b_k=\varphi(a_k)$ for $k=0,1,\dots,n-1$, the numbers

$$\alpha_k=\frac{b_{k+1}-b_k}{a_{k+1}-a_k} \quad \text{for } k=0,1,\dots,n-2,$$

verify $\alpha_0<\alpha_1<\dots<\alpha_{n-2}<0$, because φ is convex and strictly decreasing.

We now consider the even, piecewise linear function $\psi\colon \mathbb{R}\to[0,\infty[$ defined in the following way. The graph of ψ on the interval $[a_k,a_{k+1}]$ for $k=0,\dots,n-2$, is the straight line connecting the points (a_k,b_k) and (a_{k+1},b_{k+1}). In order to define ψ on the interval $[a_{n-1},\infty[$ we choose $\alpha_{n-1}\in]\alpha_{n-2},0[$ and put

$$a_n=a_{n-1}-\frac{b_{n-1}}{\alpha_{n-1}}.$$

On the interval $[a_{n-1},a_n]$ the graph of ψ is defined to be the straight line connecting (a_{n-1},b_{n-1}) and $(a_n,0)$, and on the interval $[a_n,\infty[$ we put $\psi=0$.

We then have

$$|\varphi(x)-\psi(x)|<\varepsilon$$

for all $x\in\mathbb{R}$, but since

$$\psi(x)=\sum_{k=1}^{n}\lambda_k\varphi_{a_k}(x),$$

where $\lambda_k=a_k(\alpha_k-\alpha_{k-1})>0$ for $k=1,2,\dots,n$ (we put $\alpha_n=0$), ψ is positive definite.

It follows that φ is uniform limit of continuous positive definite functions, hence positive definite. □

5.5. As an application of Polya's theorem we find that the function

$$p_\alpha(x)=e^{-|x|^\alpha}$$

is continuous and positive definite for $\alpha\in[0,1]$.

Since

$$\lim_{x\to0}\frac{1-p_\alpha(x)}{x^2}=0$$

for $\alpha>2$, it follows from Proposition 5.3 that p_α is not positive definite for $\alpha>2$. We shall later see that p_α is positive definite also for $\alpha\in]1,2]$, cf. 10.5. The function p_α is for $\alpha=1,2$ included in the above table.

5.6. Exercise. A continuous function $\varphi\colon \mathbb{R}\to[0,\infty[$ verifies the conditions (i) and (ii) of Polya's theorem if and only if there exist a positive bounded measure μ on $]0,\infty[$ and a number $\alpha\geqq 0$ such that

$$\varphi(x)=\int_0^\infty \varphi_a(x)\,d\mu(a)+\alpha \qquad \text{for } x\in\mathbb{R}. \tag{3}$$

In (3) α and μ are uniquely determined as $\alpha=\lim_{x\to\infty}\varphi(x)$ and $\mu=L\varphi$ in the distribution sense, where L is the differential operator $Lf(x)=x\,D^2 f(x)$ on the interval $]0,\infty[$.

The representation (3) yields a new proof of Polya's theorem.

5.7. Further information about positive definite functions on $\mathbb{R}$ can be found in Lukacs [1] and Feller [1].

§ 6. Periodicity

After having introduced the notion of periodicity for measures and functions on the LCA-group G we study periodicity properties in terms of the Fourier transformation.

For a periodic function $f\colon G\to\mathbb{C}$ with all elements in a closed subgroup H as periods we define the quotient function $\dot{f}\colon G/H\to\mathbb{C}$ by $\dot{f}(\dot{x})=f(x)$, where $\dot{x}\in G/H$ is the equivalence class containing x.

For a measure μ on G, which is periodic with all elements in H as periods, we define, extending the above notion for functions, a quotient measure $\dot{\mu}$ on G/H.

Finally it is proved, that the Haar measure on a closed subgroup H of G is positive definite on G, and that its Fourier transform is a Haar measure on the orthogonal complement $H^\perp$.

Let G be a LCA-group with dual group Γ and let ω_G be a fixed Haar measure on G.

6.1. Definition. A measure μ on G is called *periodic* with $p\in G$ as *period* if $\tau_p\mu=\mu$, i.e. if $\varepsilon_p*\mu=\mu$.

The set of periods for μ is denoted per (μ).

It is easy to see that the set of periods per (μ) for an arbitrary measure μ is a closed subgroup of G. We call it the *periodicity group* for μ.

Let f be a continuous function on G. An element $p\in G$ is a period for the measure $f\omega_G$ if and only if $\tau_p f=\varepsilon_p*f=f$, i.e. if and only if

$$f(x-p)=f(x) \qquad \text{for all } x\in G.$$

The set per $(f\omega_G)$ will be written per (f) and elements of per (f) will be called *periods* for f.

6.2. Proposition. *Every non-zero bounded measure μ on G has compact periodicity group.*

Proof. There exists a function $f \in C_c(G)$ such that $g = \mu * f$ is different from zero. For a point $x_0 \in G$ such that $g(x_0) \neq 0$ we have

$$x_0 + \operatorname{per}(\mu) \subseteq \{x \in G \mid g(x) = g(x_0)\},$$

and the set on the right-hand side is compact since $g \in C_0(G)$. Hence $\operatorname{per}(\mu)$ is compact. □

6.3. Proposition. *For every bounded measure μ on G we have*
(i) $\operatorname{per}(\mu) = [\operatorname{supp}(\hat{\mu})]^{\perp}$,
(ii) $\operatorname{per}(\hat{\mu}) = [\operatorname{supp}(\mu)]^{\perp}$.

Proof. (i) By the injectivity of the Fourier transformation an element $p \in G$ satisfies $\varepsilon_p * \mu = \mu$ if and only if

$$\overline{(p, \gamma)}\, \hat{\mu}(\gamma) = \hat{\mu}(\gamma) \quad \text{for all } \gamma \in \Gamma,$$

and this is equivalent to

$$(p, \gamma) = 1 \quad \text{for all } \gamma \in \operatorname{supp}(\hat{\mu}),$$

i.e. $p \in [\operatorname{supp}(\hat{\mu})]^{\perp}$.
(ii) The formula

$$\varepsilon_\gamma * \hat{\mu} = (\gamma \mu)\hat{} \quad \text{for } \gamma \in \Gamma,$$

shows that $\gamma \in \operatorname{per}(\hat{\mu})$ if and only if $\gamma \mu = \mu$, which is equivalent to γ being 1 on the support of μ, i.e. $\gamma \in [\operatorname{supp}(\mu)]^{\perp}$. □

6.4. Proposition. *For a continuous and positive definite function φ on G we have*

$$\operatorname{per}(\varphi) = \{x \in G \mid \varphi(x) = \varphi(0)\} = [\operatorname{supp}(\sigma)]^{\perp},$$

where σ is the measure on Γ associated with φ.

Proof. It follows by Proposition 6.3 (ii) that

$$\operatorname{per}(\varphi) = [\operatorname{supp}(\check{\sigma})]^{\perp} = [\operatorname{supp}(\sigma)]^{\perp}.$$

A period x for φ obviously satisfies $\varphi(x) = \varphi(0)$, and conversely if $\varphi(x) = \varphi(0)$ for a $x \in G$ it follows from the inequality (5) of 3.4 (we may assume that $\varphi(0) = 1$) that $\varphi(x + y) = \varphi(y)$ for all $y \in G$. This implies that $x \in \operatorname{per}(\varphi)$. □

6.5. Proposition. *For a positive definite measure μ on G we have*
(i) $\operatorname{per}(\mu) = [\operatorname{supp}(\mathscr{F}\mu)]^{\perp}$,
(ii) $\operatorname{per}(\mathscr{F}\mu) = [\operatorname{supp}(\mu)]^{\perp}$.

Proof. (i) For all $f, g \in C_c(G)$ and all $x \in G$ we have

$$\mu * f * \tilde{g}(x) = \int (x, \gamma)\, \hat{f}(\gamma)\, \overline{\hat{g}(\gamma)}\, d\mathscr{F}\mu(\gamma), \tag{1}$$

and replacing f by $\varepsilon_p * f$ for a $p \in G$ we get

$$\mu * \varepsilon_p * f * \tilde{g}(x) = \int (x, \gamma)\, \overline{(p, \gamma)}\, \hat{f}(\gamma)\, \overline{\hat{g}(\gamma)}\, d\mathscr{F}\mu(\gamma). \tag{2}$$

If p is a period for μ, (1) and (2) show that the bounded measures $\hat{f}\,\bar{\hat{g}}\,\mathscr{F}\mu$ and $\overline{(p,\cdot)}\,\hat{f}\,\bar{\hat{g}}\,\mathscr{F}\mu$ are equal, because they have the same co-Fourier transform. It follows by Proposition 2.4 that $(p,\gamma)=1$ for all $\gamma\in\operatorname{supp}(\mathscr{F}\mu)$, i.e. $p\in[\operatorname{supp}(\mathscr{F}\mu)]^{\perp}$.

If conversely $p\in G$ satisfies $(p,\gamma)=1$ for all $\gamma\in\operatorname{supp}(\mathscr{F}\mu)$, it follows by (1) and (2) that

$$\mu * f * \tilde{g}(x)=\mu * \varepsilon_p * f * \tilde{g}(x)$$

for all $f, g\in C_c(G)$ and all $x\in G$.

Letting g and f run through an approximate unit we get $\mu=\varepsilon_p * \mu$, i.e. $p\in\operatorname{per}(\mu)$.

(ii) is contained in Corollary 4.11. □

6.6. Let H be a closed subgroup of G and let ω_H be a fixed Haar measure on H. The canonical homomorphism of G onto G/H is denoted π.

For every continuous function $f\colon G\to\mathbb{C}$ for which $\operatorname{per}(f)\supseteq H$, there exists a uniquely determined continuous function $\dot{f}\colon G/H\to\mathbb{C}$ such that $\dot{f}\circ\pi=f$, i.e. such that

$$\dot{f}(\dot{x})=f(x) \tag{3}$$

for all $x\in G$, where $\dot{x}=\pi(x)$.

We call $\dot{f}$ the *quotient function* of f over H.

6.7. For $f\in C_c(G)$ the function $\omega_H * f$ on G (ω_H is considered as a measure on G) is continuous and periodic with all $p\in H$ as periods because

$$\varepsilon_p *(\omega_H * f)=(\varepsilon_p * \omega_H) * f=\omega_H * f.$$

The quotient function $(\omega_H * f)^{\boldsymbol{\cdot}}\colon G/H\to\mathbb{C}$ of $\omega_H * f$ over H will be denoted $\sigma(f)$, so that for all $x\in G$

$$\sigma(f)(\dot{x})=\omega_H * f(x)=\int f(x+y)\,d\omega_H(y). \tag{4}$$

If $\sigma(f)(\dot{x})\neq 0$ there exists $y\in H$ such that $f(x+y)\neq 0$, and hence

$$\pi(x)=\pi(x+y)\in\pi(\operatorname{supp}(f)).$$

This shows that

$$\operatorname{supp}(\sigma(f))\subseteq\pi(\operatorname{supp}(f)), \tag{5}$$

and in particular $\sigma(f)\in C_c(G/H)$.

6.8. Lemma. *For every compact subset $C\subseteq G/H$ there exists a compact subset $K\subseteq G$ such that $\pi(K)=C$.*

Proof. Let U be a compact neighbourhood of 0 in G. Since π is an open and continuous mapping, $\pi(U)$ is a compact neighbourhood of 0 in G/H. There exist finitely many points $x_1, \ldots, x_n \in G$ such that

$$C \subseteq \bigcup_{i=1}^{n} (\pi(x_i) + \pi(U)).$$

The set

$$K = \left(\bigcup_{i=1}^{n} (x_i + U)\right) \cap \pi^{-1}(C)$$

is compact and $\pi(K) = C$. □

6.9. Proposition. *The mapping $\sigma\colon C_c(G) \to C_c(G/H)$ determined by (4) is continuous, positive, linear and surjective in the following strong sense: For every $h \in C_c^+(G/H)$ there exists $f \in C_c^+(G)$ such that $\sigma(f) = h$.*

For $a \in G$ and $f \in C_c(G)$ we have

$$\sigma(\tau_a f) = \tau_{\pi(a)} \sigma(f). \tag{6}$$

Proof. It is clear that σ is positive and linear. Let K be a compact subset of G and let $(f_n)_{n \in \mathbb{N}}$ be a sequence of functions from $C_c(G)$ converging uniformly to a function $f \in C_c(G)$ such that the support of f_n is contained in K for every n. Then $\omega_H * f_n$ tends to $\omega_H * f$ uniformly on compact sets, in particular

$$\sup_{x \in K} |\omega_H * f_n(x) - \omega_H * f(x)|$$

tends to zero. However,

$$\sup_{x \in K} |\omega_H * f_n(x) - \omega_H * f(x)| = \sup_{\dot{x} \in \pi(K)} |\sigma(f_n)(\dot{x}) - \sigma(f)(\dot{x})|,$$

so by (5) $\sigma(f_n)$ tends to $\sigma(f)$ uniformly. This proves that σ is continuous.

For $h \in C_c^+(G/H)$ we choose according to Lemma 6.8 a compact set $K \subseteq G$ such that $\pi(K) = \operatorname{supp}(h)$. We next choose a function $\varphi \in C_c^+(G)$ such that $\varphi = 1$ on K.

The function

$$f(x) = \begin{cases} \dfrac{\varphi(x)\, h(\pi(x))}{\omega_H * \varphi(x)} & \text{if } \omega_H * \varphi(x) \neq 0, \\[2ex] 0 & \text{if } \omega_H * \varphi(x) = 0, \end{cases}$$

is easily seen to be in $C_c^+(G)$, because $\omega_H * \varphi(x) = 0$ implies $\pi(x) \notin \pi(K) = \operatorname{supp}(h)$.

We find that

$$\omega_H * f(x) = \begin{cases} h(\pi(x)) & \text{if } \omega_H * \varphi(x) \neq 0, \\ 0 & \text{if } \omega_H * \varphi(x) = 0, \end{cases}$$

so that $\sigma(f) = h$.

For $f \in C_c(G)$ and $a, x \in G$ we have

$$\tau_{\pi(a)}\sigma(f)(\pi(x)) = \sigma(f)(\pi(x-a)) = \omega_H * f(x-a)$$
$$= \omega_H * \tau_a f(x) = \sigma(\tau_a f)(\pi(x)),$$

so that

$$\sigma(\tau_a f) = \tau_{\pi(a)}\sigma(f). \quad \square$$

6.10. The transpose of σ is the linear map $\sigma^*: M(G/H) \to M(G)$ defined by

$$\langle \sigma^*(\mu), f\rangle = \langle \mu, \sigma(f)\rangle \quad \text{for } \mu \in M(G/H) \text{ and } f \in C_c(G). \tag{7}$$

The periodicity group for $\sigma^*(\mu)$, where $\mu \in M(G/H)$, contains H. In fact, using (6) we have for $a \in H$ and $f \in C_c(G)$

$$\langle \tau_a \sigma^*(\mu), f\rangle = \langle \mu, \sigma(\tau_{-a} f)\rangle = \langle \mu, \tau_{-\pi(a)}\sigma(f)\rangle$$
$$= \langle \mu, \sigma(f)\rangle = \langle \sigma^*(\mu), f\rangle,$$

because $\pi(a)$ is zero in G/H.

Let $\omega_{G/H}$ be a Haar measure on G/H. Then $\sigma^*(\omega_{G/H})$ is a non-zero positive measure on G. Furthermore it is a Haar measure on G, because for all $a \in G$ and $f \in C_c(G)$ we have by (6)

$$\langle \tau_a \sigma^*(\omega_{G/H}), f\rangle = \langle \omega_{G/H}, \sigma(\tau_{-a} f)\rangle = \langle \omega_{G/H}, \tau_{-\pi(a)}\sigma(f)\rangle$$
$$= \langle \omega_{G/H}, \sigma(f)\rangle = \langle \sigma^*(\omega_{G/H}), f\rangle.$$

This proves the following result:

6.11. Proposition. *There exists a Haar measure $\omega_{G/H}$ on G/H with the property that $\sigma^*(\omega_{G/H}) = \omega_G$, i.e. such that*

$$\int_G f\, d\omega_G = \int_{G/H} \sigma(f)\, d\omega_{G/H} \quad \textit{for all } f \in C_c(G).$$

The formula in Proposition 6.11 is sometimes written in the following unprecise form

$$\int_G f(x)\, d\omega_G(x) = \int_{G/H} \Big(\int_H f(x+h)\, d\omega_H(h)\Big)\, d\omega_{G/H}(\dot{x}).$$

The measure $\omega_{G/H}$ is thus a "quotient" between ω_G and ω_H, and we shall now see how it is possible, via the mapping σ^*, to define the quotient over H of a periodic measure.

6.12. Lemma. *Let μ be a measure on G whose periodicity group contains H, and let f be a function in $C_c(G)$ such that $\sigma(f) = 0$. Then $\langle \mu, f\rangle = 0$.*

Proof. For every $\varphi\in C_c(G)$ we have

$$\omega_H*(f(\mu*\varphi))=(\omega_H*f)(\mu*\varphi),$$

because $\mu*\varphi$ is periodic with all $p\in H$ as periods. It follows that

$$\sigma(f(\mu*\varphi))=\sigma(f)(\mu*\varphi)^{\cdot}=0,$$

and by Proposition 6.11 we therefore get

$$\int f(\mu*\varphi)\,d\omega_G=\int\sigma(f(\mu*\varphi))\,d\omega_{G/H}=0,$$

where $\omega_{G/H}$ is the Haar measure on G/H such that $\sigma^*(\omega_{G/H})=\omega_G$.

This proves that $\langle\mu,f*\check{\varphi}\rangle=0$ for every $\varphi\in C_c(G)$. Letting φ run through an approximate unit we finally get $\langle\mu,f\rangle=0$. □

6.13. Proposition. *Let μ be a positive measure on G whose periodicity group contains H. There exists a uniquely determined measure $\dot{\mu}$ on G/H such that $\sigma^*(\dot{\mu})=\mu$. The measure $\dot{\mu}$ is positive.*

Proof. For every $h\in C_c(G/H)$ there exists by Proposition 6.9 a function $f\in C_c(G)$ such that $\sigma(f)=h$. Any measure $\dot{\mu}$ on G/H satisfying $\sigma^*(\dot{\mu})=\mu$ therefore verifies $\langle\dot{\mu},h\rangle=\langle\mu,f\rangle$, and it follows by Lemma 6.12 that a mapping $\dot{\mu}\colon C_c(G/H)\to\mathbb{C}$ is well-defined by putting

$$\langle\dot{\mu},h\rangle=\langle\mu,f\rangle\qquad\text{for } h\in C_c(G/H),$$

where $f\in C_c(G)$ satisfies $\sigma(f)=h$. It is obvious that this mapping $\dot{\mu}$ is a positive linear form on $C_c(G/H)$ i.e. $\dot{\mu}$ is a positive measure on G/H, and by definition we have that $\sigma^*(\dot{\mu})=\mu$. □

6.14. Corollary. *The mapping σ^* is an isomorphism of $M(G/H)$ onto the subspace of measures μ on G, which are periodic with all elements $p\in H$ as periods.*

Proof. As remarked in 6.10, σ^* maps $M(G/H)$ into the set of measures μ on G for which $\operatorname{per}(\mu)\supseteq H$, and as the transpose of the surjective mapping σ, we have that σ^* is injective.

Let μ be a real measure on G for which $\operatorname{per}(\mu)\supseteq H$. Then μ can be decomposed as $\mu=\mu^+-\mu^-$, where $\mu^+=\max(\mu,0)$, $\mu^-=\max(-\mu,0)$, and it is easily seen that μ^+ and μ^- are periodic with all $p\in H$ as periods. It follows by Proposition 6.13 that there exists a real measure $\dot{\mu}$ on G/H such that $\sigma^*(\dot{\mu})=\mu$.

Let finally μ be a complex measure on G such that $\operatorname{per}(\mu)\supseteq H$. The real and imaginary parts of μ are also periodic with all $p\in H$ as periods. It follows that there exists a measure $\dot{\mu}$ on G/H such that $\sigma^*(\dot{\mu})=\mu$. □

6.15. Let μ be a measure on G such that $\operatorname{per}(\mu)\supseteq H$. The uniquely determined measure $\dot{\mu}$ on G/H such that $\sigma^*(\dot{\mu})=\mu$ is called the *quotient measure* of μ over H. It is determined by the formula

$$\langle\dot{\mu},\sigma(f)\rangle=\langle\mu,f\rangle\qquad\text{for } f\in C_c(G).\tag{8}$$

6.16. Remark. It should be noted that the mapping σ^* and the quotient measure $\dot{\mu}$ of a periodic measure μ depend on the choice of a Haar measure ω_H on H.

6.17. Exercise. Let f be a continuous function on G such that $\operatorname{per}(f) \supseteq H$. Then $(f\,\omega_G)^{\dot{}} = \dot{f}\,\dot{\omega}_G$ where $\dot{\omega}_G = \omega_{G/H}$.

6.18. Exercise. Let K be a compact subgroup of G and let ω_K be the Haar measure on K with total mass 1. For any measure μ on G such that $\operatorname{per}(\mu) \supseteq K$ the quotient measure $\dot{\mu}$ of μ over K is equal to the image measure $\pi(\mu)$ under the canonical mapping $\pi\colon G \to G/K$. Let μ and ν be bounded measures on G such that $\operatorname{per}(\mu) \supseteq K$ and $\operatorname{per}(\nu) \supseteq K$. Then $\operatorname{per}(\mu * \nu) \supseteq K$ and $(\mu * \nu)^{\dot{}} = \dot{\mu} * \dot{\nu}$.

The following theorem may be considered as a generalisation of the Poisson summation formula, cf. 6.21 and 6.22.

6.19. Theorem. *Any Haar measure ω_H for a closed subgroup H of G is a positive definite measure on G. The Fourier transform $\mathscr{F}_G\,\omega_H$ is a Haar measure on the subgroup $H^\perp$ of Γ.*

Proof. For $f \in C_c(G)$ we have by Proposition 6.11

$$\langle \omega_H, f * \tilde{f} \rangle = \int (\omega_H * \bar{f})\, f\, d\omega_G = \int |\sigma(f)|^2\, d\omega_{G/H},$$

which shows that ω_H is positive definite. By Proposition 6.5 it follows that the Fourier transform $\mathscr{F}_G\,\omega_H$ is a positive measure on $H^\perp$, which has all elements $p \in H^\perp$ as periods. The measure $\mathscr{F}_G\,\omega_H$ being different from zero is thus a Haar measure on $H^\perp$. □

6.20. Remark. Let K be a compact subgroup of G and denote by ω_K the normalized Haar measure on K.

The Fourier transform of the bounded measure ω_K is the function

$$\hat{\omega}_K(\gamma) = 1_{K^\perp}(\gamma), \tag{9}$$

which is a continuous function on Γ, and in particular the subgroup $K^\perp$ of Γ is open.

The measure $1_{K^\perp}\,\omega_\Gamma$ is a Haar measure on $K^\perp$, namely the Fourier transform of the positive definite measure ω_K.

The equation (9) can be seen in the following way: It is clear that

$$\hat{\omega}_K(\gamma) = 1 \qquad \text{for } \gamma \in K^\perp.$$

For $\gamma \notin K^\perp$ there exists $k \in K$ such that $(k, \gamma) \neq 1$ and we find

$$\hat{\omega}_K(\gamma) = \int \overline{(x, \gamma)}\, d\omega_K(x) = \int \overline{(x+k, \gamma)}\, d\omega_K(x) = \overline{(k, \gamma)}\,\hat{\omega}_K(\gamma).$$

This implies that $\hat{\omega}_K(\gamma) = 0$.

6.21. Remark. For $\gamma \in H^{\perp}$ we have $\mathrm{per}(\gamma) \supseteq H$, cf. Proposition 6.4, and the quotient function $\dot{\gamma}$ of γ over H is a continuous character on G/H. The mapping $\gamma \mapsto \dot{\gamma}$ is an isomorphism and a homeomorphism of $H^{\perp}$ onto $\widehat{G/H}$. The inverse mapping is $\hat{\pi}$, cf. 2.11.

For $f \in C_c(G)$ and $\gamma \in H^{\perp}$ we have $\sigma(f)\dot{\gamma} = \sigma(\gamma f)$ and it is then easy to see that

$$\mathscr{F}_G f(\gamma) = \mathscr{F}_{G/H}\,\sigma(f)(\dot{\gamma}). \tag{10}$$

Since ω_H is symmetric, we have by Theorem 4.7, cf. the proof of 6.19, that

$$\langle \omega_H, f * \tilde{f} \rangle = \int |\mathscr{F}_G f|^2 \, d\mathscr{F}_G \omega_H = \int |\sigma(f)|^2 \, d\omega_{G/H}. \tag{11}$$

When considering $H^{\perp}$ as the dual group of G/H, (10) and (11) show that $\mathscr{F}_G \omega_H$ is the dual Haar measure of $\omega_{G/H}$. Putting $\omega_{H^{\perp}} = \mathscr{F}_G \omega_H$, the formula (11) can be written

$$\int_H \varphi \, d\omega_H = \int_{H^{\perp}} \mathscr{F}_G \varphi \, d\omega_{H^{\perp}}, \tag{12}$$

where $\varphi = f * \tilde{f}$ for $f \in C_c(G)$. By polarization (12) is also valid for functions of the form $\varphi = f * g$, where $f, g \in C_c(G)$.

6.22. Example. (The Poisson summation formula.) Let $G = \mathbb{R}$ and $H = \mathbb{Z}$. As Haar measure on $\mathbb{Z}$ we use

$$\omega_{\mathbb{Z}} = \sum_{n \in \mathbb{Z}} \varepsilon_n.$$

The orthogonal complement $\mathbb{Z}^{\perp}$ of $\mathbb{Z}$ is (via the identification $\hat{\mathbb{R}} \approx \mathbb{R}$, cf. 5.1)

$$\mathbb{Z}^{\perp} = 2\pi\mathbb{Z}.$$

The quotient group $\mathbb{R}/\mathbb{Z}$ is compact and for the function $f \in C_c^+(\mathbb{R})$ defined by

$$f(x) = \begin{cases} 1 - |x| & \text{for } |x| \leq 1, \\ 0 & \text{for } |x| > 1, \end{cases}$$

we find $\omega_{\mathbb{Z}} * f(x) = 1$ for all $x \in \mathbb{R}$, i.e. $\sigma(f)(\dot{x}) = 1$ for all $\dot{x} \in \mathbb{R}/\mathbb{Z}$. It follows by Proposition 6.11 that the Haar measure $\omega_{\mathbb{R}/\mathbb{Z}}$ such that $\sigma^*(\omega_{\mathbb{R}/\mathbb{Z}}) = \omega_{\mathbb{R}}$ has total mass one.

The dual Haar measure to $\omega_{\mathbb{R}/\mathbb{Z}}$ on $(\mathbb{R}/\mathbb{Z})\hat{} \approx 2\pi\mathbb{Z}$ is therefore the measure

$$\omega_{2\pi\mathbb{Z}} = \sum_{n \in \mathbb{Z}} \varepsilon_{2\pi n},$$

and it follows by Theorem 6.19 and Remark 6.21 that

$$\mathscr{F}_{\mathbb{R}}\Big(\sum_{n\in\mathbb{Z}} \varepsilon_n\Big)=\sum_{n\in\mathbb{Z}} \varepsilon_{2\pi n}.$$

In particular we have

$$\sum_{n\in\mathbb{Z}} \varphi(n)=\sum_{n\in\mathbb{Z}} \hat{\varphi}(2\pi n)$$

for all functions $\varphi=f*g$, where $f, g\in C_c(\mathbb{R})$, and this is the *Poisson summation formula.*

6.23. Remark. For further information about quotients of periodic measures, we refer to Choquet and Deny [3].

Chapter II. Negative Definite Functions and Semigroups

§ 7. Negative Definite Functions

Let G be a LCA-group with dual group Γ. In this paragraph shall we introduce the negative definite functions. After giving some equivalent formulations of the notion of negative definiteness, elementary properties and simple examples of negative definite functions are discussed. Many of the considerations are of an algebraic nature and do not involve the topology of the group.

7.1. Definition. A function $\psi\colon \Gamma \to \mathbb{C}$ is called *negative definite* if for all natural numbers n and all n-tuples $(\gamma_1, \ldots, \gamma_n)$ of elements from Γ, the $n \times n$ matrix

$$(\psi(\gamma_i)+\overline{\psi(\gamma_j)}-\psi(\gamma_i-\gamma_j))$$

is positive hermitian, i.e. if

$$\sum_{i,j=1}^{n} (\psi(\gamma_i)+\overline{\psi(\gamma_j)}-\psi(\gamma_i-\gamma_j))\, c_i \bar{c}_j \geqq 0 \tag{1}$$

for any n-tuple $(c_1, \ldots, c_n) \in \mathbb{C}^n$.

We denote by $N(\Gamma)$ the set of negative definite functions on Γ and by $CN(\Gamma)$ the set of continuous negative definite functions on Γ.

7.2. Let $\psi \in N(\Gamma)$. It is clear that

$$\psi(0) \geqq 0. \tag{2}$$

For every $\gamma \in \Gamma$ the matrix

$$\begin{pmatrix} \psi(\gamma)+\overline{\psi(\gamma)}-\psi(0) & \psi(\gamma)+\overline{\psi(0)}-\psi(\gamma) \\ \psi(0)+\overline{\psi(\gamma)}-\psi(-\gamma) & \psi(0)+\overline{\psi(0)}-\psi(0) \end{pmatrix}$$

is positive hermitian, and it follows that $\psi(-\gamma)=\overline{\psi(\gamma)}$ and (by considering the determinant) that $\operatorname{Re}\psi(\gamma) \geqq \psi(0)$, i.e.

$$\psi = \tilde{\psi} \tag{3}$$

and

$$\operatorname{Re}\psi \geqq \psi(0). \tag{4}$$

7.3. Exercise. Let $\psi: \Gamma \to \mathbb{R}$ satisfy $\psi = \tilde{\psi}$. If (1) holds for all $\gamma_1, \ldots, \gamma_n \in \Gamma$ and all $c_1, \ldots, c_n \in \mathbb{Z}$, then $\psi \in N(\Gamma)$.

It is easy to establish the following facts about the sets $N(\Gamma)$ and $CN(\Gamma)$.

7.4. Proposition. (i) *The set $N(\Gamma)$ is a convex cone which is closed in the topology of pointwise convergence on Γ.*

(ii) *For $\psi \in N(\Gamma)$ we have $\bar{\psi} \in N(\Gamma)$ and* $\operatorname{Re} \psi \in N(\Gamma)$.

(iii) *The non-negative constant functions belong to $N(\Gamma)$.*

(iv) *The set $CN(\Gamma)$ is a convex cone which is closed in the topology of compact convergence.*

7.5. Proposition. *A function $\psi: \Gamma \to \mathbb{C}$ is negative definite if and only if the following three conditions are satisfied:*

(i) $\psi(0) \geqq 0$,

(ii) $\psi = \tilde{\psi}$,

(iii) *For all $n \in \mathbb{N}$, $\gamma_1, \ldots, \gamma_n \in \Gamma$* and $c_1, \ldots, c_n \in \mathbb{C}$,

$$\sum_{i=1}^{n} c_i = 0 \quad \textit{implies} \quad \sum_{i,j=1}^{n} \psi(\gamma_i - \gamma_j) c_i \bar{c}_j \leqq 0.$$

Proof. Suppose first that $\psi \in N(\Gamma)$. It is clear that (i) and (ii) are satisfied. Let $n \in \mathbb{N}$, $\gamma_1, \ldots, \gamma_n \in \Gamma$ and $c_1, \ldots, c_n \in \mathbb{C}$ be such that $\sum_{i=1}^{n} c_i = 0$. Then we find

$$\begin{aligned} 0 &\leqq \sum_{i,j=1}^{n} \left(\psi(\gamma_i) + \overline{\psi(\gamma_j)} - \psi(\gamma_i - \gamma_j)\right) c_i \bar{c}_j \\ &= \sum_{j=1}^{n} \bar{c}_j \left(\sum_{i=1}^{n} \psi(\gamma_i) c_i\right) + \sum_{i=1}^{n} c_i \left(\sum_{j=1}^{n} \overline{\psi(\gamma_j)} \bar{c}_j\right) - \sum_{i,j=1}^{n} \psi(\gamma_i - \gamma_j) c_i \bar{c}_j \\ &= -\sum_{i,j=1}^{n} \psi(\gamma_i - \gamma_j) c_i \bar{c}_j. \end{aligned}$$

Conversely, suppose that ψ satisfies (i)–(iii), and consider $\gamma_1, \ldots, \gamma_n \in \Gamma$ and $c_1, \ldots, c_n \in \mathbb{C}$. For the $(n+1)$-tuples $(0, \gamma_1, \ldots, \gamma_n)$ and $(c, c_1, \ldots, c_n)$, where $c = -\sum_{i=1}^{n} c_i$, we get by (iii)

$$\psi(0)|c|^2 + \sum_{i=1}^{n} \psi(\gamma_i) c_i \bar{c} + \sum_{j=1}^{n} \psi(-\gamma_j) c \bar{c}_j + \sum_{i,j=1}^{n} \psi(\gamma_i - \gamma_j) c_i \bar{c}_j \leqq 0,$$

hence, using (i) and (ii), that

$$\sum_{i,j=1}^{n} \left(\psi(\gamma_i) + \overline{\psi(\gamma_j)} - \psi(\gamma_i - \gamma_j)\right) c_i \bar{c}_j \geqq \psi(0)|c|^2 \geqq 0. \quad \square$$

7.6. Corollary. *Let ψ be a negative definite function on Γ. The function*

$$\gamma \mapsto \psi(\gamma) - \psi(0) \tag{5}$$

is negative definite.

Proof. Let $\gamma_1, \ldots, \gamma_n \in \Gamma$ and $c_1, \ldots, c_n \in \mathbb{C}$ be given satisfying $\sum_{i=1}^n c_i = 0$. Then we find

$$\sum_{i,j=1}^{n} (\psi(\gamma_i - \gamma_j) - \psi(0))\, c_i \bar{c}_j = \sum_{i,j=1}^{n} \psi(\gamma_i - \gamma_j)\, c_i \bar{c}_j \leqq 0$$

because $\psi \in N(\Gamma)$, and since the function (5) clearly satisfies (i) and (ii) of Proposition 7.5, it is negative definite. □

7.7. Corollary. *Let φ be a positive definite function on Γ. The function*

$$\gamma \mapsto \varphi(0) - \varphi(\gamma) \tag{6}$$

is negative definite.

Proof. Let $\gamma_1, \ldots, \gamma_n \in \Gamma$ and $c_1, \ldots, c_n \in \mathbb{C}$ be given satisfying $\sum_{i=1}^n c_i = 0$. Then we find

$$\sum_{i,j=1}^{n} (\varphi(0) - \varphi(\gamma_i - \gamma_j))\, c_i \bar{c}_j = -\sum_{i,j=1}^{n} \varphi(\gamma_i - \gamma_j)\, c_i \bar{c}_j \leqq 0$$

because $\varphi \in P(\Gamma)$, and since the function (6) clearly satisfies (i) and (ii) of Proposition 7.5, it is negative definite. □

We now come to the most important result of the paragraph, which establishes the connection between negative definiteness and positive definiteness. For a "continuous" version of this result see Corollary 8.4.

7.8. Theorem (Schoenberg). *A function $\psi: \Gamma \to \mathbb{C}$ is negative definite if and only if the following two conditions are satisfied:*

(i) $\psi(0) \geqq 0$,

(ii) *The function $\gamma \mapsto \exp(-t\psi(\gamma))$ is positive definite for all $t > 0$.*

Proof. Suppose first that $\psi \in N(\Gamma)$. Condition (i) is clearly satisfied. Let

$$\gamma_1, \ldots, \gamma_n \in \Gamma.$$

The matrix

$$(\psi(\gamma_i) + \overline{\psi(\gamma_j)} - \psi(\gamma_i - \gamma_j))$$

being positive hermitian, it follows (cf. 3.2) that the matrix

$$(\exp(\psi(\gamma_i) + \overline{\psi(\gamma_j)} - \psi(\gamma_i - \gamma_j)))$$

is positive hermitian. For $c_1, \ldots, c_n \in \mathbb{C}$ we then find

$$\begin{aligned}
&\sum_{i,j=1}^{n} \exp(-\psi(\gamma_i - \gamma_j))\, c_i \bar{c}_j \\
&\quad = \sum_{i,j=1}^{n} \exp(\psi(\gamma_i) + \overline{\psi(\gamma_j)} - \psi(\gamma_i - \gamma_j)) \exp(-\psi(\gamma_i)) \exp(-\overline{\psi(\gamma_j)})\, c_i \bar{c}_j \\
&\quad = \sum_{i,j=1}^{n} \exp(\psi(\gamma_i) + \overline{\psi(\gamma_j)} - \psi(\gamma_i - \gamma_j))\, c'_i \bar{c}'_j \geqq 0,
\end{aligned}$$

where $c_i'=\exp(-\psi(\gamma_i))\,c_i\in\mathbb{C}$. It follows that $\exp(-\psi)$ is positive definite, and condition (ii) is therefore satisfied as $t\psi$ is negative definite for all $t>0$.

Conversely, suppose that ψ satisfies (i) and (ii). By (i) we have $\exp(-t\psi(0))\leqq 1$ for all $t>0$, and it follows that the function

$$\gamma\mapsto\frac{1}{t}(1-\exp(-t\psi(\gamma)))$$

is negative definite for all $t>0$ (cf. 7.7). Furthermore for all $\gamma\in\Gamma$

$$\psi(\gamma)=\lim_{t\to 0}\frac{1}{t}(1-\exp(-t\psi(\gamma))),$$

and Proposition 7.4 now gives that $\psi\in N(\Gamma)$. □

7.9. Corollary. *Let $\psi\colon\Gamma\to\mathbb{C}$ be negative definite and suppose that $\psi(0)>0$. Then $1/\psi$ is positive definite.*

Proof. By Schoenberg's Theorem the function

$$\gamma\mapsto\exp(-t\psi(\gamma))$$

is positive definite for all $t>0$. Furthermore

$$|\exp(-t\psi(\gamma))|\leqq\exp(-t\psi(0))$$

for all $t>0$ and $\gamma\in\Gamma$, and it follows that

$$\frac{1}{\psi(\gamma)}=\int_0^\infty\exp(-t\psi(\gamma))\,dt\qquad\text{for } \gamma\in\Gamma,$$

which shows that $1/\psi\in P(\Gamma)$. □

7.10. Exercise. Let $\psi\in N(\Gamma)$. For all $\alpha>0$ and $\beta\geqq 0$ the function $\psi(\alpha+\beta\psi)^{-1}$ is negative definite.

7.11. Proposition. *A function $\psi\colon\Gamma\to\mathbb{C}$ is negative definite if and only if there exists a sequence $(\psi_n)_{n\in\mathbb{N}}$ of functions $\psi_n\colon\Gamma\to\mathbb{C}$ of the form*

$$\psi_n=a_n+\varphi_n(0)-\varphi_n,$$

where $a_n\geqq 0$ and $\varphi_n\colon\Gamma\to\mathbb{C}$ is positive definite, such that

$$\lim_{n\to\infty}\psi_n=\psi\qquad\textit{pointwise on } \Gamma.$$

Proof. For the "only if"-part we consider for $n \in \mathbb{N}$ the function $\varphi_n: \Gamma \to \mathbb{C}$ defined by

$$\varphi_n(\gamma) = n \exp\left[-\frac{1}{n}(\psi(\gamma) - \psi(0))\right].$$

By Theorem 7.8 $\varphi_n \in P(\Gamma)$ for all $n \in \mathbb{N}$, and putting

$$a_n = \psi(0) \quad \text{and} \quad \psi_n(\gamma) = a_n + \varphi_n(0) - \varphi_n(\gamma)$$

for $n \in \mathbb{N}$, we find, using the power series expansion for exp, that

$$\psi(\gamma) - \psi_n(\gamma) = \frac{1}{n}\left[\frac{(\psi(\gamma) - \psi(0))^2}{2!} - \frac{(\psi(\gamma) - \psi(0))^3}{n \cdot 3!} + \cdots\right]$$

for $\gamma \in \Gamma$. It follows that

$$|\psi(\gamma) - \psi_n(\gamma)| \leqq \frac{1}{n} \exp(|\psi(\gamma) - \psi(0)|) \qquad \text{for } \gamma \in \Gamma,$$

and the conclusion follows.

The "if"-part is clear by Propositions 7.4 and 7.7. □

7.12. Remark. Let $\psi: \Gamma \to \mathbb{C}$ be a continuous and negative definite function. There exists a sequence of functions $\psi_n: \Gamma \to \mathbb{C}$ of the form

$$\psi_n = a_n + \varphi_n(0) - \varphi_n,$$

where $a_n \geqq 0$ and $\varphi_n: \Gamma \to \mathbb{C}$ is continuous and positive definite, such that

$$\lim_{n \to \infty} \psi_n = \psi$$

uniformly on compact subsets of Γ.

The constants a_n and the functions φ_n and ψ_n from the proof of Proposition 7.11 have these properties.

It is clear that a function ψ of the form

$$\psi = a + \varphi(0) - \varphi, \tag{7}$$

where $a \geqq 0$ and φ is continuous and positive definite, is bounded, continuous and negative definite. The next proposition shows conversely, that a bounded, continuous and negative definite function is of the form (7). This result is taken from Harzallah [1].

7.13. Proposition. *Let $\psi: \Gamma \to \mathbb{C}$ be a bounded and continuous negative definite function. Then there exists a constant $m > 0$ such that $m - \psi$ is positive definite.*

Proof. We may (and do) suppose that $\psi(0)=0$. For $t>0$ we consider the negative definite function

$$\psi_t=\frac{1}{t}[1-\exp(-t\psi)].$$

The function ψ_t can be written

$$\psi_t(\gamma)=\int_G (1-\overline{(x,\gamma)})\,da_t(x) \quad \text{for } \gamma\in\Gamma,$$

where a_t is a positive bounded measure on G, with total mass $a_t(G)\leqq 1/t$, such that $a_t(\{0\})=0$. Using the power series expansion for exp we find

$$|\psi-\psi_t|\leqq t\exp|\psi| \quad \text{for } t>0,$$

and it follows that

$$\lim_{t\to 0}\psi_t=\psi \quad \text{uniformly on } \Gamma. \tag{8}$$

It is clear that ψ_t is of the form

$$\psi_t=m_t-\varphi_t$$

for "many" couples (m_t,φ_t), where $m_t\geqq 0$ and $\varphi_t\in CP(\Gamma)$, and choosing for every $t>0$ an appropriate pair (m_t,φ_t), the constant m will be obtained as the limit of m_t as $t\to 0$. For this purpose let $\check{V}(0)$ be a basis for the system of neighbourhoods of 0 in G, and choose for every $V\in\check{V}(0)$ a continuous, positive definite function f_V on G such that

$$\operatorname{supp}(f_V)\subseteq V, \quad 0\leqq f_V\leqq 1 \quad \text{and} \quad f_V(0)=1.$$

(For $V\in\check{V}(0)$ one chooses $g_V\in C_c^+(G)$ such that $\operatorname{supp}(g_V)-\operatorname{supp}(g_V)\subseteq V$ and $\int g_V^2(x)\,dx=1$. Then $f_V=g_V*\check{g}_V$ has these properties.)

The positive bounded measure on Γ associated with f_V is denoted σ_V, and it is symmetric and has total mass 1. For $t>0$ and $V\in\check{V}(0)$ we find

$$\langle\sigma_V,\psi_t\rangle=\int_G (1-f_V(x))\,da_t(x),$$

and it follows that

$$\lim_{\check{V}(0)}\langle\sigma_V,\psi_t\rangle=a_t(G\setminus\{0\})=a_t(G).$$

By (8) there exists for every $\varepsilon>0$ a $t_0>0$ such that

$$|\psi-\psi_t|<\varepsilon \quad \text{on } \Gamma \text{ for all } t\in]0,t_0[,$$

and since $\langle\sigma_V,\psi\rangle$ is real we find

$$a_t(G)-\varepsilon\leqq\liminf_{\check{V}(0)}\langle\sigma_V,\psi\rangle\leqq\limsup_{\check{V}(0)}\langle\sigma_V,\psi\rangle\leqq a_t(G)+\varepsilon$$

for all $t \in]0, t_0[$. It follows that the limit

$$m = \lim_{\check{V}(0)} \langle \sigma_V, \psi \rangle$$

exists and furthermore that

$$m = \lim_{t \to 0} a_t(G).$$

Finally, as

$$a_t(G) - \psi_t(\gamma) = \int \overline{(x, \gamma)}\, da_t(x) \qquad \text{for } \gamma \in \Gamma$$

and

$$\lim_{t \to 0} (a_t(G) - \psi_t) = m - \psi \qquad \text{pointwise,}$$

the function $m - \psi$ is positive definite as limit of a net of positive definite functions. □

7.14. Exercise. Let $\psi\colon \Gamma \to \mathbb{C}$ be a negative definite function and let $\alpha \in]0, 1[$. The function $\psi^\alpha\colon \Gamma \to \mathbb{C}$ defined by

$$\psi^\alpha(\gamma) = \begin{cases} \exp\left[\alpha(\log|\psi(\gamma)| + i \arg \psi(\gamma))\right] & \text{if } \psi(\gamma) \neq 0, \\ 0 & \text{if } \psi(\gamma) = 0, \end{cases}$$

$\left(\text{where } \arg z \text{ is the argument belonging to } \left[-\frac{\pi}{2}, \frac{\pi}{2}\right] \text{ for } z \in \mathbb{C} \text{ with } \operatorname{Re} z \geqq 0 \text{ and } z \neq 0\right)$ is negative definite.

If furthermore ψ is continuous, then $\psi^\alpha\colon \Gamma \to \mathbb{C}$ is a continuous, negative definite function.

Hint: Suppose first that ψ is of the form $\psi = c + \varphi(0) - \varphi$ where $c > 0$ and φ is positive definite, and use the power series expansion

$$(1 - x)^\alpha = 1 - \sum_{n=1}^{\infty} (-1)^{n-1} \binom{\alpha}{n} x^n \qquad \text{for } |x| < 1.$$

The above result will also be obtained as a special case of the general theory of §9.

7.15. Proposition. *For a negative definite function ψ on Γ the function $\sqrt{|\psi|}$ is subadditive, i.e.*

$$\sqrt{|\psi(\gamma + \delta)|} \leqq \sqrt{|\psi(\gamma)|} + \sqrt{|\psi(\delta)|} \qquad \textit{for all } \gamma, \delta \in \Gamma.$$

Proof. Let $\gamma, \delta \in \Gamma$. Since the matrix

$$\begin{pmatrix} \psi(\gamma) + \overline{\psi(\gamma)} - \psi(0) & \psi(\gamma) + \overline{\psi(\delta)} - \psi(\gamma - \delta) \\ \psi(\delta) + \overline{\psi(\gamma)} - \psi(\delta - \gamma) & \psi(\delta) + \overline{\psi(\delta)} - \psi(0) \end{pmatrix}$$

has non-negative determinant we find, using $\psi=\tilde{\psi}$,

$$|\psi(\gamma)+\overline{\psi(\delta)}-\psi(\gamma-\delta)|^2 \leqq (2\operatorname{Re}\psi(\gamma)-\psi(0))\,(2\operatorname{Re}\psi(\delta)-\psi(0))$$
$$\leqq 4\,|\psi(\gamma)|\,|\psi(\delta)|$$

hence, again using $\psi=\tilde{\psi}$, that

$$|\psi(\gamma+\delta)| \leqq \left(\sqrt{|\psi(\gamma)|}+\sqrt{|\psi(\delta)|}\right)^2. \quad \square$$

7.16. Corollary. *Let $\psi\in N(\mathbb{R}^n)$ be locally bounded. Then there exists $c>0$, such that*

$$|\psi(x)| \leqq c\|x\|^2 \quad \textit{for } \|x\| \geqq 1,$$

i.e. $\psi\in\mathcal{O}(\|x\|^2)$ for $\|x\|\to\infty$.

Proof. For all $y\in\mathbb{R}^n$ and $n\in\mathbb{N}$ we get by Proposition 7.15 that

$$\sqrt{|\psi(ny)|} \leqq n\sqrt{|\psi(y)|},$$

or

$$|\psi(x)| \leqq n^2\left|\psi\left(\frac{x}{n}\right)\right| \quad \text{for } n\in\mathbb{N} \text{ and } x\in\mathbb{R}^n.$$

Putting $c=\sup_{\|z\|\leqq 2}|\psi(z)|$, there exists for each $x\in\mathbb{R}^n$ such that $\|x\|\geqq 1$, an $n_0\in\mathbb{N}$ satisfying $\|x\|\in[n_0, n_0+1[$, and it follows that

$$|\psi(x)| \leqq n_0^2\left|\psi\left(\frac{x}{n_0}\right)\right| \leqq c\,\|x\|^2. \quad \square$$

7.17. Exercise. For a negative definite function ψ on Γ the set

$$\{\gamma\in\Gamma \mid \psi(\gamma)=\psi(0)\}$$

is a subgroup of Γ. Cf. 8.27.

We shall now give some general examples of negative definite functions.

7.18. Definition. A function $q\colon \Gamma\to\mathbb{R}$ is called a *quadratic form*, if it satisfies the equation

$$2\,q(\gamma)+2\,q(\delta)=q(\gamma+\delta)+q(\gamma-\delta) \quad \text{for all } \gamma,\delta\in\Gamma.$$

It is easy to see that a quadratic form q satisfies

$$q(0)=0,$$
$$q(\gamma)=q(-\gamma) \quad \text{for all } \gamma\in\Gamma,$$
$$q(n\gamma)=n^2 q(\gamma) \quad \text{for all } \gamma\in\Gamma \text{ and } n\in\mathbb{N}.$$

7.19. Proposition. *A non-negative quadratic form q on Γ is negative definite.*

Proof. We define a mapping $B\colon \Gamma\times\Gamma\to\mathbb{R}$ by

$$B(\gamma,\delta)=q(\gamma)+q(\delta)-q(\gamma-\delta),$$

and B is a non-negative, symmetric and bi-additive form, i.e.

$$B(\gamma+\delta,\chi)=B(\gamma,\chi)+B(\delta,\chi)\quad \text{for } \gamma,\delta,\chi\in\Gamma, \tag{9}$$
$$B(\gamma,\delta+\chi)=B(\gamma,\delta)+B(\gamma,\chi)\quad \text{for } \gamma,\delta,\chi\in\Gamma, \tag{10}$$
$$B(\gamma,\delta)=B(\delta,\gamma)\quad \text{for } \gamma,\delta\in\Gamma, \tag{11}$$
$$B(\gamma,\gamma)\geqq 0\quad \text{for } \gamma\in\Gamma. \tag{12}$$

Equation (9) follows by the simple computation:

$$\begin{aligned}
&q(\gamma)+q(\chi)-q(\gamma-\chi)+q(\delta)+q(\chi)-q(\delta-\chi)\\
&=q(\gamma)+q(\delta)+2q(\chi)-\frac{1}{2}[q(\gamma-\chi+\delta-\chi)+q(\gamma-\chi-(\delta-\chi))]\\
&=q(\gamma)+q(\delta)-\frac{1}{2}q(\gamma-\delta)+2q(\chi)-\frac{1}{2}[2q(\gamma+\delta-\chi)+2q(\chi)-q(\gamma+\delta)]\\
&=\frac{1}{2}q(\gamma+\delta)+2q(\chi)-q(\gamma+\delta-\chi)-q(\chi)+\frac{1}{2}q(\gamma+\delta)\\
&=q(\gamma+\delta)+q(\chi)-q(\gamma+\delta-\chi).
\end{aligned}$$

Equations (11) and (12) are immediate, and (10) follows from (9) and (11). For $\gamma_1,\dots,\gamma_n\in\Gamma$ and $c_1,\dots,c_n\in\mathbb{Z}$ we find

$$\sum_{i,j=1}^{n}[q(\gamma_i)+\overline{q(\gamma_j)}-q(\gamma_i-\gamma_j)]c_i\bar{c}_j=B(\gamma,\gamma)\geqq 0,$$

where $\gamma=c_1\gamma_1+\dots+c_n\gamma_n$, and Exercise 7.3 now shows that $q\in N(\Gamma)$. □

7.20. Proposition. *Let l be a real function on Γ and define $\psi\colon\Gamma\to\mathbb{C}$ by $\psi(\gamma)=il(\gamma)$. The function ψ is negative definite if and only if l is a homomorphism of Γ into $\mathbb{R}$.*

Proof. Suppose that l is a homomorphism. Then for any n-tuple $(\gamma_1,\dots,\gamma_n)$ of elements from Γ, the matrix

$$(\psi(\gamma_i)+\overline{\psi(\gamma_j)}-\psi(\gamma_i-\gamma_j))$$

is the zero-matrix, and it follows that $\psi\in N(\Gamma)$.

Suppose next that $\psi\in N(\Gamma)$. For all $t>0$ the function $\varphi_t\colon \gamma\mapsto\exp(-t\psi(\gamma))$ is positive definite and has absolute value 1. It follows by 3.5 that $\varphi_t\colon\Gamma\to\mathbb{T}$ is a character on Γ, and consequently for $\gamma,\delta\in\Gamma$ and $t>0$

$$\exp(-til(\gamma+\delta))=\exp(-ti(l(\gamma)+l(\delta))),$$

and this implies

$$l(\gamma+\delta)=l(\gamma)+l(\delta). \quad \square$$

7.21. The negative definite functions have been introduced by Schoenberg [1] in connection with isometric imbeddings of groups into Hilbert spaces. Many of the above results seem to have been discovered by Beurling (unpublished) and used in the work of Beurling and Deny [1], where the translation invariant Dirichlet spaces on G are characterized by means of the real continuous negative definite functions on Γ. The present exposition is inspired by lectures of Deny [6], which develop the results of Beurling and Deny. The next paragraph is devoted to the important relation between convolution semigroups and continuous negative definite functions.

§ 8. Convolution Semigroups

Let G be a LCA-group with dual group Γ. We shall consider families of positive bounded measures on G, which induce strongly continuous contraction semigroups on various Banach spaces of functions on G, cf. § 12. The first main result is the one-to-one correspondence between such families – the convolution semigroups on G – and continuous, negative definite functions on Γ. The resolvent for a convolution semigroup is introduced, and we find conditions on an "abstract resolvent", which are necessary and sufficient for this resolvent to be the resolvent for a convolution semigroup. Finally it is shown, that a convolution semigroup on G is concentrated on a σ-compact subgroup of G.

8.1. Definition. A family $(\mu_t)_{t>0}$ of positive bounded measures on G with the properties

(i) $\mu_t(G)\leqq 1$ for $t>0$,

(ii) $\mu_t * \mu_s = \mu_{t+s}$ for $t, s>0$,

(iii) $\lim_{t\to 0} \mu_t = \varepsilon_0$ vaguely,

is called a (vaguely continuous) *convolution semigroup* on G.

8.2. Proposition. *Let* $(\mu_t)_{t>0}$ *be a convolution semigroup on* G *and put* $\mu_0=\varepsilon_0$. *The mapping* $t\mapsto\mu_t$ *of* $[0,\infty[$ *into* $M_b^+(G)$ *is continuous in the Bernoulli topology.*

Proof. For $f\in C_c^+(G)$ satisfying $0\leqq f\leqq 1$ and $f(0)=1$ we find using conditions (iii) and (i) above

$$1=f(0)=\lim_{t\to 0}\langle\mu_t, f\rangle\leqq\liminf_{t\to 0}\mu_t(G)\leqq\limsup_{t\to 0}\mu_t(G)\leqq 1,$$

and this shows by Proposition 1.4 that

$$\lim_{t\to 0}\mu_t=\varepsilon_0 \quad \text{in the Bernoulli topology.} \tag{1}$$

For $t, t_0>0$ and $\gamma\in\Gamma$ we find

$$|\hat{\mu}_t(\gamma)-\hat{\mu}_{t_0}(\gamma)|\leqq|\hat{\mu}_{|t-t_0|}(\gamma)-1|,$$

and since the right-hand side, by (1), tends to zero uniformly on compact subsets of Γ (cf. 3.13), this shows that

$$\lim_{t\to t_0}\mu_t=\mu_{t_0}$$

in the Bernoulli topology. □

8.3. Theorem. *There is a one-to-one correspondence between convolution semigroups $(\mu_t)_{t>0}$ on G and continuous, negative definite functions ψ on Γ. More precisely: If $(\mu_t)_{t>0}$ is a convolution semigroup on G, then there exists a uniquely determined continuous, negative definite function ψ on Γ such that*

$$\hat{\mu}_t(\gamma)=\exp(-t\psi(\gamma))\qquad\text{for } t>0 \text{ and } \gamma\in\Gamma. \tag{2}$$

Conversely, given a continuous, negative definite function ψ on Γ, then (2) *determines a convolution semigroup $(\mu_t)_{t>0}$ on G.*

Proof. Let $(\mu_t)_{t>0}$ be a convolution semigroup on G, and consider, for a fixed $\gamma\in\Gamma$, the function φ_γ: $]0,\infty[\to\mathbb{C}$ defined by

$$\varphi_\gamma(t)=\hat{\mu}_t(\gamma)\qquad\text{for } t>0.$$

The function φ_γ is continuous by Proposition 8.2, and it clearly satisfies

$$\varphi_\gamma(s+t)=\varphi_\gamma(s)\varphi_\gamma(t)\quad\text{and}\quad\lim_{t\to 0}\varphi_\gamma(t)=1,$$

and there exists consequently a uniquely determined complex number $\psi(\gamma)$ such that

$$\varphi_\gamma(t)=\exp(-t\psi(\gamma))\qquad\text{for } t>0.$$

The function $\gamma\mapsto\psi(\gamma)$ defined in this way has the properties a) $\psi(0)\geqq 0$ and b) $\gamma\mapsto\exp(-t\psi(\gamma))=\hat{\mu}_t(\gamma)$ is continuous and positive definite for all $t>0$. By Schoenberg's theorem (cf. 7.8) it follows that ψ is negative definite. Moreover, the measure ρ, defined by

$$\langle\rho,f\rangle=\int_0^\infty e^{-t}\langle\mu_t,f\rangle\,dt\qquad\text{for } f\in C_c(G),$$

is positive and bounded with total mass $\leqq 1$, and we find

$$\hat{\rho}(\gamma)=\int_0^\infty e^{-t}\hat{\mu}_t(\gamma)\,dt=\int_0^\infty\exp(-t(1+\psi(\gamma)))\,dt=\frac{1}{1+\psi(\gamma)}\qquad\text{for } \gamma\in\Gamma,$$

and as $\hat{\rho}$ is continuous, this implies that $\psi\in CN(\Gamma)$.

Conversely, let $\psi \in CN(\Gamma)$. For every $t>0$ the function

$$\gamma \mapsto \exp(-t\psi(\gamma))$$

is continuous and positive definite (cf. 7.8), and there exists consequently a positive, bounded measure μ_t on G such that

$$\hat{\mu}_t(\gamma) = \exp(-t\psi(\gamma)) \quad \text{for } \gamma \in \Gamma.$$

We shall see that the family $(\mu_t)_{t>0}$ is a convolution semigroup. Since $\psi(0) \geqq 0$, it is clear that $\mu_t(G) = \hat{\mu}_t(0) = \exp(-t\psi(0)) \leqq 1$ for all $t>0$. Furthermore for $t, s>0$ and $\gamma \in \Gamma$

$$\hat{\mu}_t(\gamma)\hat{\mu}_s(\gamma) = \exp(-t\psi(\gamma))\exp(-s\psi(\gamma)) = \exp(-(t+s)\psi(\gamma)) = \hat{\mu}_{t+s}(\gamma),$$

and it follows that $\mu_t * \mu_s = \mu_{t+s}$. Finally, as ψ is continuous, hence bounded on compact sets, we have

$$\lim_{t\to 0} \hat{\mu}_t(\gamma) = \lim_{t \to 0} \exp(-t\psi(\gamma)) = 1$$

uniformly on compact subsets of Γ, and it follows by Theorem 3.13 that

$$\lim_{t \to 0} \mu_t = \varepsilon_0$$

in the Bernoulli topology. □

From the proof of Theorem 8.3 we get the following "continuous" version of Schoenberg's theorem (cf. 7.8).

8.4. Corollary. *A function $\psi: \Gamma \to \mathbb{C}$ is continuous and negative definite if and only if the following two conditions are satisfied:*

(i) $\psi(0) \geqq 0$.

(ii) *The function $\gamma \mapsto \exp(-t\psi(\gamma))$ is continuous and positive definite for all $t>0$.*

8.5. Definition. If the convolution semigroup $(\mu_t)_{t>0}$ on G and the continuous, negative definite function ψ on Γ correspond to each other by the correspondence of Theorem 8.3, we will say that $(\mu_t)_{t>0}$ and ψ are *associated.*

8.6. Corollary. *The total masses of the measures in a convolution semigroup $(\mu_t)_{t>0}$ on G are given by*

$$\mu_t(G) = \exp(-t\psi(0)) \quad \textit{for } t>0,$$

where ψ is the continuous negative definite function on Γ associated with $(\mu_t)_{t>0}$, and in particular $(\mu_t)_{t>0}$ consists of probability measures if and only if $\psi(0)=0$.

8.7. Let $(\mu_t)_{t>0}$ be a convolution semigroup on G with associated continuous negative definite function ψ on Γ. The family $(\check{\mu}_t)_{t>0}$ of reflected measures is a convolution semigroup on G with associated continuous negative definite function $\bar{\psi}$ on Γ. We will say that $(\mu_t)_{t>0}$ is *symmetric* if $\mu_t=\check{\mu}_t$ for $t>0$, and this is the case if and only if ψ is real. The measures μ_t in a symmetric convolution semigroup $(\mu_t)_{t>0}$ on G are positive definite because $\mu_t=\mu_{t/2}*\check{\mu}_{t/2}$ for $t>0$.

8.8. Exercise. Let $(\mu_t^{(i)})_{t>0}$, $i=1,2$, be convolution semigroups on G and let ψ_i, $i=1,2$, be the associated continuous negative definite functions on Γ. For $\alpha,\beta>0$ the function $\alpha\psi_1+\beta\psi_2$ is continuous and negative definite and the associated convolution semigroup $(\mu_t)_{t>0}$ is given by

$$\mu_t=\mu_{\alpha t}^{(1)}*\mu_{\beta t}^{(2)} \quad \text{for } t>0.$$

8.9. Exercise. Let G_1 and G_2 be LCA-groups with dual groups Γ_1 and Γ_2 and consider convolution semigroups $(\mu_t^{(i)})_{t>0}$ on G_i, $i=1,2$, with associated negative definite functions ψ_i on Γ_i, $i=1,2$.

The function $\psi\colon \Gamma_1\times\Gamma_2\to\mathbb{C}$ defined by

$$\psi((\gamma_1,\gamma_2))=\psi_1(\gamma_1)+\psi_2(\gamma_2) \quad \text{for } (\gamma_1,\gamma_2)\in\Gamma_1\times\Gamma_2$$

belongs to $CN(\Gamma_1\times\Gamma_2)$ and the associated convolution semigroup $(\mu_t)_{t>0}$ on $G_1\times G_2$ (cf. 2.12) is given by

$$\mu_t=\mu_t^{(1)}\otimes\mu_t^{(2)} \quad \text{for } t>0.$$

8.10. Let $\mu\in M_b^+(G)$ and $\alpha\geqq\mu(G)$. A positive bounded measure μ_t on G is defined for $t>0$ by

$$\mu_t=e^{-t\alpha}\exp(t\mu)=e^{-t\alpha}\sum_{n=0}^{\infty}\frac{t^n}{n!}\mu^n,$$

where μ^n denotes the n-fold convolution of μ with itself, and we put $\mu^0=\varepsilon_0$.

The Fourier transform of μ_t is

$$\hat{\mu}_t(\gamma)=\exp(-t(\alpha-\hat{\mu}(\gamma))) \quad \text{for } \gamma\in\Gamma,$$

and it follows that $(\mu_t)_{t>0}$ is a convolution semigroup with associated negative definite function $\alpha-\hat{\mu}$.

For a positive bounded measure μ on G such that $\mu(G)\leqq 1$, the convolution semigroup $(\mu_t)_{t>0}$ on G defined by

$$\mu_t=e^{-t}\exp(t\mu) \quad \text{for } t>0,$$

will be called the *convolution semigroup determined by* μ.

8.11. A family $(\mu_t)_{t>0}$ of measures on G of the form $\mu_t=\varepsilon_{x(t)}$ for $t>0$, where $x\colon [0,\infty[\to G$ is a continuous mapping satisfying

$$x(s+t)=x(s)+x(t) \quad \text{for } s,t\geqq 0,$$

is called a *translation semigroup*. It is easy to see that a translation semigroup is a convolution semigroup.

Let $(\varepsilon_{x(t)})_{t>0}$ be a translation semigroup on G, and define a mapping $\varphi\colon \mathbb{R}\to G$ by

$$\varphi(s)=\begin{cases} x(s) & \text{for } s\geqq 0,\\ -x(-s) & \text{for } s<0.\end{cases}$$

It is clear that φ is continuous, and it is easy to see that φ is a homomorphism. Consider the dual homomorphism $l=\hat{\varphi}\colon \Gamma\to\mathbb{R}$. For the Fourier transform of the measure $\varepsilon_{x(t)}$ we find $(\gamma\in\Gamma)$,

$$\begin{aligned}\hat{\varepsilon}_{x(t)}(\gamma)&=\langle\varepsilon_{x(t)},\bar{\gamma}\rangle=\bar{\gamma}(x(t))=\bar{\gamma}(\varphi(t))\\ &=\overline{(\varphi(t),\gamma)}=\overline{(t,\hat{\varphi}(\gamma))}=\exp(-til(\gamma)).\end{aligned}$$

Conversely, let $l\colon \Gamma\to\mathbb{R}$ be a continuous homomorphism and define $x=\hat{l}\,|\,[0,\infty[$. It is clear that $(\varepsilon_{x(t)})_{t>0}$ is a translation semigroup on G and that

$$\hat{\varepsilon}_{x(t)}(\gamma)=\exp(-til(\gamma)) \quad \text{for } t>0 \text{ and } \gamma\in\Gamma. \tag{3}$$

There is thus a one-to-one correspondence between translation semigroups on G and real, continuous homomorphisms on Γ. The correspondence is given via the Fourier transformation by (3). Let $l\colon \Gamma\to\mathbb{R}$ be the continuous homomorphism corresponding to the translation semigroup $(\varepsilon_{x(t)})_{t>0}$. The continuous, negative definite function associated with the convolution semigroup $(\varepsilon_{x(t)})_{t>0}$ is given by $\gamma\mapsto il(\gamma)$. Cf. Proposition 7.20.

8.12. Let $(\mu_t)_{t>0}$ be a convolution semigroup on G and $\psi\colon \Gamma\to\mathbb{C}$ the associated continuous, negative definite function.

For $\lambda>0$ we define a positive measure ρ_λ on G by

$$\langle\rho_\lambda,f\rangle=\int_0^\infty e^{-\lambda t}\langle\mu_t,f\rangle\,dt \tag{4}$$

for $f\in C_c(G)$, and the equation (4) holds for any non-negative, bounded continuous function f on G; in particular

$$\rho_\lambda(G)=\int_0^\infty e^{-\lambda t}\mu_t(G)\,dt=\int_0^\infty e^{-t(\lambda+\psi(0))}\,dt=\frac{1}{\lambda+\psi(0)}.$$

The measure ρ_λ is consequently bounded and of total mass

$$\rho_\lambda(G) \leqq \frac{1}{\lambda}. \tag{5}$$

Furthermore, the measure $\lambda\rho_\lambda$ is a probability measure if and only if $\psi(0)=0$, i.e. if and only if $(\mu_t)_{t>0}$ consists of probability measures (cf. 8.6).

The Fourier transform of ρ_λ is given by the following formula

$$\hat{\rho}_\lambda(\gamma) = \int_0^\infty e^{-\lambda t}\langle\mu_t, \bar{\gamma}\rangle\, dt = \int_0^\infty e^{-t(\lambda+\psi(\gamma))}\, dt = \frac{1}{\lambda+\psi(\gamma)} \quad \text{for } \gamma\in\Gamma. \tag{6}$$

It is now easy to see that the family $(\rho_\lambda)_{\lambda>0}$ of positive bounded measures on G satisfies the relation

$$\rho_\lambda - \rho_\mu = (\mu-\lambda)\rho_\lambda * \rho_\mu \quad \text{for } \lambda, \mu > 0. \tag{7}$$

In fact, the Fourier transform of the left-hand side of (7) is

$$\mathscr{F}_G(\rho_\lambda - \rho_\mu)(\gamma) = \frac{1}{\lambda+\psi(\gamma)} - \frac{1}{\mu+\psi(\gamma)} = \frac{\mu-\lambda}{(\lambda+\psi(\gamma))(\mu+\psi(\gamma))} \quad \text{for } \gamma\in\Gamma,$$

which is equal to the Fourier transform of the right-hand side of (7), and (7) now follows since $\mathscr{F}_G$ is injective.

The family $(\rho_\lambda)_{\lambda>0}$ is called the *resolvent for the convolution semigroup* $(\mu_t)_{t>0}$, and (7) will be referred to as the *resolvent equation*.

A convolution semigroup $(\mu_t)_{t>0}$ is uniquely determined by its resolvent $(\rho_\lambda)_{\lambda>0}$, since by (6) the resolvent determines the associated negative definite function ψ uniquely. The semigroup can even be obtained by a limit procedure from the resolvent, as shown by the following proposition, where the first statement is a special case of the Hille-Yosida theorem for semigroups of operators, cf. 11.19, and the second is a special case of the Post-Widder inversion formula for the Laplace transformation.

8.13. Proposition. *Let* $(\rho_\lambda)_{\lambda>0}$ *be the resolvent for the convolution semigroup* $(\mu_t)_{t>0}$ *on G. Then*

$$\mu_t = \lim_{\lambda\to\infty} \mu_t^\lambda \quad \textit{and} \quad \mu_t = \lim_{k\to\infty}\left[\frac{k}{t}\rho_{k/t}\right]^k \quad \textit{for } t>0, \tag{8}$$

in the Bernoulli topology, where $(\mu_t^\lambda)_{t>0}$ *for* $\lambda>0$ *is the convolution semigroup on G defined by*

$$\mu_t^\lambda = e^{-\lambda t}\exp(t\lambda^2\rho_\lambda) \quad \textit{for } t>0.$$

Proof. Let $\psi: \Gamma \to \mathbb{C}$ be the continuous negative definite function associated with $(\mu_t)_{t>0}$. By 8.10 the family $(\mu_t^\lambda)_{t>0}$ is a convolution semigroup on G and we find

$$\mathscr{F}_G(\mu_t^\lambda)(\gamma) = e^{-\lambda t} \exp\left(\frac{t\lambda^2}{\lambda+\psi(\gamma)}\right) = \exp\left(-t\frac{\lambda\psi(\gamma)}{\lambda+\psi(\gamma)}\right) \qquad \text{for } \gamma\in\Gamma \text{ and } t, \lambda>0,$$

and consequently

$$\lim_{\lambda\to\infty} \mathscr{F}_G(\mu_t^\lambda)(\gamma) = \exp(-t\psi(\gamma)) = \mathscr{F}_G(\mu_t)(\gamma) \qquad \text{for } t>0,$$

pointwise (and even uniformly over compact sets) on Γ, and the first equation of (8) follows by Theorem 3.14.

For $k\in\mathbb{N}$, $t>0$ and $\gamma\in\Gamma$ we have

$$\mathscr{F}_G\left(\left(\frac{k}{t}\rho_{k/t}\right)^k\right)(\gamma) = \left(1+\frac{t\psi(\gamma)}{k}\right)^{-k}$$

and consequently

$$\lim_{k\to\infty} \mathscr{F}_G\left(\left(\frac{k}{t}\rho_{k/t}\right)^k\right)(\gamma) = \exp(-t\psi(\gamma)) = \mathscr{F}_G(\mu_t)(\gamma) \qquad \text{for } t>0,$$

pointwise on Γ, and the second equation in (8) follows by Theorem 3.14. □

In order to characterize the families $(\rho_\lambda)_{\lambda>0}$ of positive bounded measures on G, which are resolvents for convolution semigroups on G, shall we now consider families of measures satisfying the equation (7), cf. Forst [2].

8.14. Definition. A family $(\rho_\lambda)_{\lambda>0}$ of positive bounded measures on G with total mass $\lambda\rho_\lambda(G)\leqq 1$ will be called a *resolvent of measures* on G, if it satisfies the *resolvent equation*

$$\rho_\lambda - \rho_\mu = (\mu-\lambda)\rho_\lambda * \rho_\mu \qquad \text{for } \lambda, \mu>0.$$

A family $(\mu_t)_{t>0}$ of positive bounded measures on G with total mass $\mu_t(G)\leqq 1$ will be called a *semigroup of measures* on G, if it satisfies

$$\mu_t * \mu_s = \mu_{t+s} \qquad \text{for } t, s>0. \tag{9}$$

Let $(\rho_\lambda)_{\lambda>0}$ be a resolvent of measures on G. If there exists a $\lambda>0$ such that $\rho_\lambda=0$, then $(\rho_\lambda)_{\lambda>0}$ is the zero family. Likewise, a semigroup of measures $(\mu_t)_{t>0}$ on G such that $\mu_t=0$ for a $t>0$, is the zero family, and in the sequel (8.15–8.23) we will always assume that $\rho_\lambda\neq 0$ for all $\lambda>0$ (resp. $\mu_t\neq 0$ for all $t>0$).

In the following sections these notions will be studied and in particular it will be shown that every resolvent of measures is the "integral" of a uniquely determined semigroup of measures.

8.15. Lemma. *Let* $(\mu_t)_{t>0}$ *be a semigroup of measures on G. The set*

$$\{\gamma\in\Gamma \mid \hat{\mu}_t(\gamma)\neq 0\},$$

which is independent of $t>0$, *is an open subgroup* H *of* Γ, *and the compact subgroup* $K=H^{\perp}$ *of* G *is the periodicity group for all the measures* μ_t, $t>0$.

If furthermore $(\mu_t)_{t>0}$ *is symmetric* (i.e. $\mu_t=\check{\mu}_t$ *for* $t>0$), *then*

$$\lim_{t\to 0}\mu_t=\omega_K$$

in the Bernoulli topology, where ω_K *is the normalized Haar measure on* K.

Proof. Let $0<t<s$ and choose $n\in\mathbb{N}$ such that $s<nt$. Putting $H_t=\{\gamma\in\Gamma\mid\hat{\mu}_t(\gamma)\neq 0\}$ for $t>0$, it is clear by (9) that

$$H_s\subseteq H_t=H_{nt}\subseteq H_s,$$

and it follows that H_t is independent of $t>0$.

We first suppose that $(\mu_t)_{t>0}$ is symmetric. For $t>0$ we then have

$$0\leqq\hat{\mu}_t(\gamma)\leqq 1 \quad \text{for } \gamma\in\Gamma, \tag{10}$$

and hence

$$\lim_{n\to\infty}\hat{\mu}_{1/n}(\gamma)=\lim_{n\to\infty}(\hat{\mu}_1(\gamma))^{1/n}=1_H(\gamma). \tag{11}$$

This shows that 1_H, as limit function for a sequence of positive definite functions, is positive definite, and in particular we get that H is a subgroup of Γ (cf. Exercise 3.8). Since $\hat{\mu}_1>0$ in a neighbourhood of 0 ($\mu_t\neq 0$ for all $t>0$) H is open (and closed), and 1_H is consequently a continuous positive definite function. Since the support of $\hat{\mu}_t$ is H, it follows by Proposition 6.3 that K is the periodicity group for all the measures μ_t, $t>0$. By (10) it is seen that the function $t\mapsto\hat{\mu}_t(\gamma)$ is decreasing for all $\gamma\in\Gamma$, and it follows by (11) that

$$\lim_{t\to 0}\hat{\mu}_t(\gamma)=1_H(\gamma) \quad \text{pointwise on } \Gamma,$$

and this shows by Theorem 3.14 that

$$\lim_{t\to 0}\mu_t=\omega_K \quad \text{in the Bernoulli topology,}$$

because $\hat{\omega}_K=1_H$, cf. Remark 6.20.

To see that $H(=H_t)$ is an open subgroup of Γ, also in the general case, we remark that

$$\{\gamma\in\Gamma\mid\hat{\mu}_t(\gamma)\neq 0\}=\{\gamma\in\Gamma\mid[\mu_t*\check{\mu}_t]^\wedge(\gamma)\neq 0\} \quad \text{for } t>0,$$

and the assertion follows by considering the symmetric semigroup of measures $(\mu_t*\check{\mu}_t)_{t>0}$ on G. □

8.16. Lemma. *Let $(\rho_\lambda)_{\lambda>0}$ be a resolvent of measures on G. The set*

$$\{\gamma\in\Gamma \mid \hat{\rho}_\lambda(\gamma)\neq 0\},$$

which is independent of $\lambda>0$, is an open subgroup H of Γ, and the compact subgroup $K=H^\perp$ of G is the periodicity group for all the measures ρ_λ, $\lambda>0$. Denoting by ω_K the normalized Haar measure on K we have

$$\lim_{\lambda\to\infty} \lambda\rho_\lambda=\omega_K$$

in the Bernoulli topology.

Proof. It is a simple consequence of the resolvent equation that the set $\{\gamma\in\Gamma \mid \hat{\rho}_\lambda(\gamma)\neq 0\}$ is independent of $\lambda>0$. We first remark that

$$\lim_{\lambda\to\infty} \lambda\hat{\rho}_\lambda(\gamma)=1_H(\gamma) \quad \text{pointwise on } \Gamma. \tag{12}$$

This is clear for $\gamma\notin H$, and for $\gamma\in H$ we find by the resolvent equation

$$\hat{\rho}_\lambda(\gamma)=\frac{\hat{\rho}_1(\gamma)}{1+(\lambda-1)\,\hat{\rho}_1(\gamma)} \quad \text{for } \lambda>0,$$

and it follows that

$$\lim_{\lambda\to\infty} \lambda\hat{\rho}_\lambda(\gamma)=1 \quad \text{for } \gamma\in H.$$

This shows that 1_H is positive definite, and the rest of the proof is analogous to the proof of Lemma 8.15. □

Let $(\mu_t)_{t>0}$ be a semigroup of measures on G. The compact subgroup K of G from Lemma 8.15 will be called the *periodicity group* for $(\mu_t)_{t>0}$. Analogously for a resolvent of measures $(\rho_\lambda)_{\lambda>0}$ on G, the subgroup K from Lemma 8.16 will be called the *periodicity group* for $(\rho_\lambda)_{\lambda>0}$.

A semigroup of measures $(\mu_t)_{t>0}$ on G is said to be *continuous* if the mapping, $t\mapsto\mu_t$ of $]0,\infty[$ into $M_b^+(G)$ is continuous in the Bernoulli topology. As in the proof of Proposition 8.2, it can be seen that $(\mu_t)_{t>0}$ is continuous if and only if

$$\lim_{t\to 0} \mu_t=\omega_K$$

in the Bernoulli topology, where ω_K is the normalized Haar measure on the periodicity group K for $(\mu_t)_{t>0}$.

By Lemma 8.15 every semigroup of measures which consists of symmetric measures is a continuous semigroup.

8.17. Corollary. *A semigroup of measures on G is a convolution semigroup on G if and only if it is a continuous semigroup whose periodicity group is $\{0\}$.*

Proof. This is immediate. □

8.18. Remark. Let $(\mu_t)_{t>0}$ be a semigroup of measures on G with periodicity group K. The family $(\dot{\mu}_t)_{t>0}$ of quotient measures of μ_t over K (cf. 6.15) is a semigroup of measures on the quotient group G/K with periodicity group $\{0\}$ (in G/K). If $(\mu_t)_{t>0}$ is a continuous semigroup, then $(\dot{\mu}_t)_{t>0}$ is also a continuous semigroup, hence a convolution semigroup on G/K.

There exist, however, semigroups of measures (e.g. on $G=\mathbb{R}$) which are *not* continuous. Every non-continuous homomorphism $x: \mathbb{R} \to G$ determines a semigroup of measures $(\mu_t)_{t>0}$ defined by $\mu_t = \varepsilon_{x(t)}$ for $t>0$, and such a semigroup of measures is not continuous.

In analogy with the situation for convolution semigroups, continuous semigroups of measures on G can be characterized in terms of continuous negative definite functions "on" the dual group Γ.

8.19. Proposition. *Let $(\mu_t)_{t>0}$ be a continuous semigroup of measures on G with periodicity group K. Then there exists a uniquely determined continuous negative definite function ψ on the open subgroup $H=K^{\perp}$ of Γ such that for $t>0$*

$$\hat{\mu}_t(\gamma) = \begin{cases} \exp(-t\psi(\gamma)) & \text{for } \gamma \in H, \\ 0 & \text{for } \gamma \notin H. \end{cases} \tag{13}$$

Conversely, if ψ is a continuous negative definite function defined on an open subgroup H of Γ, then (13) *defines a family $(\mu_t)_{t>0}$ of positive bounded measures on G, and $(\mu_t)_{t>0}$ is a continuous semigroup of measures on G with periodicity group $K=H^{\perp}$.*

Proof. The proof is analogous to the proof of Theorem 8.3 (where $K=\{0\}$ and $H=\Gamma$). The only new difficulty consists in seeing, in the construction of the semigroup $(\mu_t)_{t>0}$, that the functions in (13) are continuous and positive definite, which is clear by Exercise 3.8, since H is an open subgroup of Γ. □

8.20. Let $(\mu_t)_{t>0}$ be a continuous semigroup of measures on G with periodicity group K, and let $\psi: H \to \mathbb{C}$ be the continuous negative definite function on $H=K^{\perp}$ associated with $(\mu_t)_{t>0}$ by Proposition 8.19. As in 8.12 we can define a family $(\rho_\lambda)_{\lambda>0}$ of positive bounded measures on G by

$$\langle \rho_\lambda, f \rangle = \int_0^\infty e^{-\lambda t} \langle \mu_t, f \rangle \, dt \qquad \text{for } f \in C_c(G) \text{ and } \lambda > 0.$$

The Fourier transform of ρ_λ is for $\lambda>0$ given by

$$\hat{\rho}_\lambda(\gamma) = \begin{cases} \dfrac{1}{\lambda + \psi(\gamma)} & \text{for } \gamma \in H, \\ 0 & \text{for } \gamma \notin H, \end{cases} \tag{14}$$

and it follows that $(\rho_\lambda)_{\lambda>0}$ is a resolvent of measures on G.

By (14) it is clear that the function ψ and hence also the continuous semigroup of measures $(\mu_t)_{t>0}$ on G is uniquely determined by $(\rho_\lambda)_{\lambda>0}$.

We shall now see that every resolvent of measures $(\rho_\lambda)_{\lambda>0}$ on G can be obtained in the above way from a continuous semigroup of measures on G.

8.21. Theorem. *Let $(\rho_\lambda)_{\lambda>0}$ be a resolvent of measures on G with periodicity group K. Then there exists a uniquely determined continuous semigroup $(\mu_t)_{t>0}$ of measures on G with K as periodicity group such that*

$$\rho_\lambda = \int_0^\infty e^{-\lambda t} \mu_t \, dt \qquad \text{vaguely for } \lambda>0. \tag{15}$$

Proof. Let $\gamma \in H = K^\perp$. It is a simple consequence of the resolvent equation that the number

$$\psi(\gamma) = \frac{1-\lambda\hat{\rho}_\lambda(\gamma)}{\hat{\rho}_\lambda(\gamma)}$$

is independent of $\lambda>0$. The function $\psi\colon H \to \mathbb{C}$ thus defined is continuous, and since

$$\lambda\hat{\rho}_\lambda(\gamma)\psi(\gamma) = \lambda(1-\lambda\hat{\rho}_\lambda(\gamma)) \qquad \text{for } \gamma \in H \text{ and } \lambda>0,$$

it follows by (12) that ψ is pointwise limit (as $\lambda\to\infty$) of negative definite functions, hence negative definite.

The continuous semigroup of measures $(\mu_t)_{t>0}$ on G associated with ψ by (13) has the desired properties, since the Fourier transform of the measure defined by the vague integral in (15) is given in terms of ψ by (14), hence equal to $\hat{\rho}_\lambda$. □

8.22. Corollary. *A resolvent of measures $(\rho_\lambda)_{\lambda>0}$ on G is resolvent for a convolution semigroup on G if and only if the periodicity group for $(\rho_\lambda)_{\lambda>0}$ is $\{0\}$, i.e. if and only if $\hat{\rho}_\lambda(\gamma) \neq 0$ for (one or) all $\lambda>0$ and all $\gamma \in \Gamma$.*

Proof. This is immediate by 8.12, Theorem 8.21 and Corollary 8.17. □

8.23. Remark. Let $(\rho_\lambda)_{\lambda>0}$ be a resolvent of measures on G with periodicity group K, and let $(\mu_t)_{t>0}$ be the uniquely determined continuous semigroup of measures on G from Theorem 8.21. The family $(\dot{\rho}_\lambda)_{\lambda>0}$ (resp. $(\dot{\mu}_t)_{t>0}$) of quotient measures of ρ_λ (resp. μ_t) over K is a resolvent of measures (resp. a convolution semigroup) on the quotient group G/K. Moreover $(\dot{\rho}_\lambda)_{\lambda>0}$ is the resolvent for the convolution semigroup $(\dot{\mu}_t)_{t>0}$.

8.24. Theorem. *Let $(\mu_t)_{t>0}$ be a convolution semigroup on G with resolvent $(\rho_\lambda)_{\lambda>0}$. For all $\lambda>0$ we have*

$$\operatorname{supp}(\rho_\lambda) = \overline{\bigcup_{t>0} \operatorname{supp}(\mu_t)},$$

and this subset of G is a σ-compact semigroup containing the neutral element. The smallest closed subgroup G_0 of G containing $\bigcup_{t>0} \operatorname{supp}(\mu_t)$ is σ-compact.

The proof is based on the following two lemmas.

8.25. Lemma. *The closure $\bar{A}$ of a σ-compact subset $A \subseteq G$ is σ-compact.*

Proof. Let $(K_n)_{n\in\mathbb{N}}$ be a sequence of compact subsets $K_n \subseteq G$ such that $A = \bigcup_{n\in\mathbb{N}} K_n$, and choose a compact, symmetric neighbourhood K of 0. Then it is easy to see that

$$\bar{A} \subseteq \bigcup_{n\in\mathbb{N}} (K_n + K)$$

and consequently

$$\bar{A} = \bigcup_{n\in\mathbb{N}} (\bar{A} \cap (K_n + K)),$$

where $\bar{A} \cap (K_n + K)$ is compact for every $n \in \mathbb{N}$. □

8.26. Lemma. *The support* $\operatorname{supp}(\mu)$ *of a positive bounded measure* μ *on* G *is* σ*-compact.*

Proof. There exists, for every $n \in \mathbb{N}$, a compact subset $K_n \subseteq G$ such that

$$\mu(G \setminus K_n) < \frac{1}{n}.$$

The set

$$A = \bigcup_{n\in\mathbb{N}} K_n$$

is σ-compact, and it is easy to see that $\operatorname{supp}(\mu) \subseteq \bar{A}$. It follows that $\operatorname{supp}(\mu)$ is σ-compact, because $\bar{A}$ is σ-compact by Lemma 8.25. □

Proof of Theorem 8.24. Let $\lambda > 0$ and $f \in C_c^+(G)$. Then $\langle \rho_\lambda, f \rangle = 0$ if and only if $\langle \mu_t, f \rangle = 0$ for all $t > 0$, and since $\lim_{t\to 0} \mu_t = \varepsilon_0$ vaguely, it follows that if $f(0) > 0$ then $\langle \rho_\lambda, f \rangle > 0$, hence

$$0 \in \operatorname{supp}(\rho_\lambda).$$

Likewise for an open set $U \subseteq G$ we have $\operatorname{supp}(\rho_\lambda) \subseteq \complement U$ if and only if

$$\operatorname{supp}(\mu_t) \subseteq \complement U$$

for all $t > 0$, and this shows that

$$\operatorname{supp}(\rho_\lambda) = \overline{\bigcup_{t>0} \operatorname{supp}(\mu_t)},$$

which by Lemma 8.26 is a σ-compact set.

For positive measures μ and ν such that $\mu * \nu$ exists we have

$$\operatorname{supp}(\mu * \nu) = \overline{\operatorname{supp}(\mu) + \operatorname{supp}(\nu)},$$

so in particular for $s, t > 0$

$$\operatorname{supp}(\mu_s) + \operatorname{supp}(\mu_t) \subseteq \operatorname{supp}(\mu_{t+s}),$$

and it follows easily that $\overline{\bigcup_{t>0} \operatorname{supp}(\mu_t)}$ is a semigroup.

The smallest closed subgroup G_0 of G containing $\bigcup_{t>0} \operatorname{supp}(\mu_t)$ is equal to

$$\overline{\operatorname{supp}(\rho_\lambda) - \operatorname{supp}(\rho_\lambda)}.$$

The set of differences

$$\operatorname{supp}(\rho_\lambda) - \operatorname{supp}(\rho_\lambda)$$

is clearly σ-compact and it follows by Lemma 8.25 that G_0 is σ-compact. □

8.27. Proposition. *Let $(\mu_t)_{t>0}$ be a convolution semigroup on G and let G_0 be the smallest closed subgroup of G containing $\bigcup_{t>0} \operatorname{supp}(\mu_t)$. The periodicity group for the associated continuous, negative definite function $\psi: \Gamma \to \mathbb{C}$ is given by*

$$\Gamma_0 = \operatorname{per}(\psi) = \{\gamma \in \Gamma \mid \psi(\gamma) = \psi(0)\},$$

and we have

$$G_0 = \Gamma_0^{\perp} \quad \text{and} \quad \Gamma_0 = G_0^{\perp}.$$

Proof. For $t>0$ we define

$$A_t = \{\gamma \in \Gamma \mid \exp(-t\psi(\gamma)) = \exp(-t\psi(0))\}.$$

It is clear that $\gamma \in \operatorname{per}(\psi)$ if and only if $\gamma \in \operatorname{per}(\exp(-t\psi))$ for all $t>0$, and it follows by Proposition 6.4, that

$$\operatorname{per}(\psi) = \bigcap_{t>0} A_t = \{\gamma \in \Gamma \mid \psi(\gamma) = \psi(0)\}.$$

Furthermore, by Proposition 6.3 and 6.4 we have

$$A_t = \operatorname{per}(\hat{\mu}_t) = [\operatorname{supp}(\mu_t)]^{\perp},$$

and therefore

$$[\operatorname{per}(\psi)]^{\perp} = [\bigcap_{t>0} [\operatorname{supp}(\mu_t)]^{\perp}]^{\perp} = [\bigcup_{t>0} \operatorname{supp}(\mu_t)]^{\perp\perp},$$

and the conclusion now follows (cf. 2.8). □

8.28. Remark. The quotient group Γ/Γ_0 is isomorphic with the dual group of G_0 (cf. 2.11) and the quotient function $\dot{\psi}$ of ψ over Γ_0 is the continuous, negative definite function associated with $(\mu_t)_{t>0}$ considered as a convolution semigroup on G_0.

8.29. Remark. The notion of convolution semigroups appears in a very natural way in the description of *translation invariant Markov processes* on a (second countable) LCA-group G. Such a Markov process X is specified by its family of *transition probabilities*, i.e. the family of numbers $P_t(x, B)$ where $P_t(x, B)$ for a "time" $t>0$, a point $x \in G$ and a Borel set $B \subseteq G$ is the probability that the process starting at "time" 0 in the point x is in the set B at "time" t.

The set of translation invariant Markov processes on G is in one-to-one correspondence with the set of convolution semigroups on G. The Markov process X (with transition probabilities $P_{\cdot}(\cdot, \cdot)$) corresponds to the convolution semigroup $(\mu_t)_{t>0}$ if and only if

$$\mu_t(B-x)=P_t(x, B) \quad \text{for } t>0,\ x\in G \quad \text{and Borel sets } B\subseteq G.$$

A detailed study of translation invariant probabilistic potential theories can be found in Port and Stone [2].

§ 9. Completely Monotone Functions and Bernstein Functions

The Laplace transformation may often be a more convenient tool than the Fourier transformation when dealing with measures on $\mathbb{R}$, which are supported by $[0, \infty[$. For convolution semigroups on $\mathbb{R}$, which are supported by $[0, \infty[$, the use of Laplace transformation instead of Fourier transformation leads to the notions of completely monotone functions and Bernstein functions instead of positive definite and negative definite functions.

After the definition of these notions the integral representations of completely monotone functions and Bernstein functions are presented.

For a continuous, negative definite function ψ on a LCA-group Γ and a Bernstein function f, the composite function $f\circ\psi$ is a continuous, negative definite function on Γ, and the corresponding convolution semigroup on the dual group G can be expressed in terms of the convolution semigroups corresponding to ψ and f.

The paragraph ends with some examples.

9.1. Definition. A C^∞-function $f\colon]0, \infty[\to\mathbb{R}$ is said to be *completely monotone*, if

$$(-1)^p D^p f\geqq 0 \quad \text{for all integers } p\geqq 0,$$

and is said to be a *Bernstein function*, if

$$f\geqq 0 \quad \text{and} \quad (-1)^p D^p f\leqq 0 \quad \text{for all integers } p\geqq 1.$$

A completely monotone function is positive, decreasing and convex. The set of completely monotone functions is a convex cone containing the positive constant functions. The product of two completely monotone functions is again completely monotone, as is easily seen by Leibniz' formula

$$D^p(fg)=\sum_{k=0}^{p}\binom{p}{k} D^k f\, D^{p-k} g.$$

A Bernstein function is positive, increasing and concave. The set of Bernstein functions is a convex cone containing the positive constant functions.

The structure of the convex cones of completely monotone functions and Bernstein functions and the relation between these cones resemble the structure of the convex cones of positive definite and negative definite functions and the relation between these cones.

The following analogue to the theorem of Schoenberg (7.8) gives an example of this resemblance, which will be explained later, cf. Remark 9.16.

9.2. Proposition. *For a function $f\colon]0,\infty[\to\mathbb{R}$ the following conditions are equivalent:*

(i) *f is a Bernstein function,*

(ii) *$f\geqq 0$ and the function $\exp(-tf)$ is completely monotone for all $t>0$.*

Proof. Suppose first that f is a Bernstein function and let $t>0$. The function $\varphi\colon]-\infty,0[\to\mathbb{R}$ defined by

$$\varphi(x)=-tf(-x) \quad \text{for } x\in]-\infty,0[$$

is of class C^∞, $\varphi\leqq 0$ and $D^p\varphi\geqq 0$ for all integers $p\geqq 1$. It is then easy to see that the function $g=\exp(\varphi)$ is of class C^∞ and that $D^p g\geqq 0$ for all integers $p\geqq 0$. This implies, that the function

$$x\mapsto g(-x)=\exp(-tf(x)),$$

defined for $x\in]0,\infty[$, is completely monotone.

Suppose next that (ii) is fulfilled. Since $\exp(-f)$ is of class C^∞, then so is f, and it is easy to justify termwise differentiation with respect to x in the series expansion

$$\exp(-tf(x))=\sum_{n=0}^{\infty}(-1)^n\frac{t^n}{n!}(f(x))^n \quad \text{for } x>0 \text{ and } t>0.$$

For $p\geqq 1$, $x>0$ and $t>0$ we find

$$0\leqq(-1)^p D^p[\exp(-tf)](x)=\sum_{n=1}^{\infty}(-1)^{n+p}\frac{t^n}{n!}D^p[f^n](x).$$

Dividing with $t>0$ and letting t tend to zero we get

$$(-1)^{1+p}D^p f(x)\geqq 0. \quad \square$$

9.3. Theorem (Bernstein). *A function $f\colon]0,\infty[\to\mathbb{R}$ is completely monotone if and only if there exists a positive measure μ on $[0,\infty[$ such that*

$$f(x)=\int_0^\infty e^{-xs}\,d\mu(s) \quad \textit{for } x>0. \tag{1}$$

For a completely monotone function f the measure μ in (1) is uniquely determined, and μ will be called the representing measure *for f.*

For the proof we refer to Meyer [1]. See also Choquet [1].

9.4. Let f be a completely monotone function with representing measure μ. Since the function f is decreasing, the limit $\lim_{x\to 0} f(x)$ exists and by (1) we get

$$\lim_{x\to 0} f(x)=\mu([0,\infty[)\leqq\infty, \tag{2}$$

and in particular $\lim_{x\to 0} f(x)<\infty$ if and only if μ is a bounded measure.

9.5. Proposition. *Let $(f_n)_{n\in\mathbb{N}}$ be a pointwise convergent sequence of completely monotone functions with representing measures $(\mu_n)_{n\in\mathbb{N}}$.*

Then the limit function

$$f(x)=\lim_{n\to\infty} f_n(x) \quad \text{for } x>0,$$

is completely monotone,

$$\lim_{n\to\infty} D^p f_n(x)=D^p f(x) \quad \text{for } x>0 \text{ and } p\in\mathbb{N},$$

and

$$\lim_{n\to\infty} \mu_n=\mu \quad \text{vaguely on } [0,\infty[,$$

where μ is the representing measure for f.

Proof. It is easy to see that $(\mu_n)_{n\in\mathbb{N}}$ is vaguely bounded. Let $(\mu_{n_p})_{p\in\mathbb{N}}$ be an arbitrary vaguely convergent subsequence with vague limit μ. For $x>0$ we clearly have

$$\int_0^\infty e^{-xs}d\mu(s)\leqq\liminf_{p\to\infty}\int_0^\infty e^{-xs}d\mu_{n_p}(s)=\lim_{p\to\infty} f_{n_p}(x)=f(x), \tag{3}$$

and for each $a>0$ we find

$$\begin{aligned} f_{n_p}(x)&=\int_0^a e^{-xs}d\mu_{n_p}(s)+\int_a^\infty e^{-\frac{1}{2}xs}e^{-\frac{1}{2}xs}d\mu_{n_p}(s)\\ &\leqq\int_0^a e^{-xs}d\mu_{n_p}(s)+e^{-\frac{1}{2}xa}f_{n_p}\left(\frac{x}{2}\right), \end{aligned}$$

and hence

$$\begin{aligned} f(x)&\leqq\limsup_{p\to\infty}\int_0^a e^{-xs}d\mu_{n_p}(s)+e^{-\frac{1}{2}xa}f\left(\frac{x}{2}\right)\\ &\leqq\int_0^a e^{-xs}d\mu(s)+e^{-\frac{1}{2}xa}f\left(\frac{x}{2}\right). \end{aligned}$$

Letting a tend to infinity we get

$$f(x)\leqq\int_0^\infty e^{-xs}d\mu(s),$$

which together with (3) shows that

$$f(x)=\int_0^\infty e^{-xs}d\mu(s). \tag{4}$$

This proves that f is completely monotone and since the measure μ is uniquely determined by (4) we get that $\lim_{n\to\infty} \mu_n = \mu$ vaguely.

Since

$$\lim_{s\to\infty} s^p e^{-as} = 0 \quad \text{for } a>0 \text{ and } p\in\mathbb{N},$$

we get as above, using that $\lim_{n\to\infty} \mu_n = \mu$ vaguely

$$\lim_{n\to\infty} D^p f_n(x) = D^p f(x) \quad \text{for } x>0 \text{ and } p\in\mathbb{N}. \quad \square$$

9.6. Proposition. *The limit function for a pointwise convergent sequence of Bernstein functions is a Bernstein function.*

Proof. This is an immediate consequence of the Propositions 9.2 and 9.5. □

9.7. Remark. The set of decreasing and convex functions f: $]0,\infty[\to\mathbb{R}$ is easily seen to be a closed and metrisable subset of $\mathbb{R}^{]0,\infty[}$ in the topology of pointwise convergence. By 9.5 we therefore have that the set of completely monotone functions is a closed and metrisable subset of $\mathbb{R}^{]0,\infty[}$ in the topology of pointwise convergence.

An analogous argument gives that the set of Bernstein functions is a closed and metrisable subset of $\mathbb{R}^{]0,\infty[}$ in the topology of pointwise convergence.

From Theorem 9.3 we can deduce an integral representation of the Bernstein functions.

9.8. Theorem. *A function f: $]0,\infty[\to\mathbb{R}$ is a Bernstein function if and only if there exist constants $a, b \geqq 0$ and a positive measure μ on $]0,\infty[$ verifying*

$$\int_0^\infty \frac{s}{1+s}\, d\mu(s) < \infty, \tag{5}$$

such that

$$f(x) = a + bx + \int_0^\infty (1-e^{-xs})\, d\mu(s) \quad \textit{for } x>0. \tag{6}$$

The triple (a, b, μ) in the representation (6) of f is uniquely determined.

Proof. Condition (5) is clearly verified if and only if

$$\int_0^1 s\, d\mu(s) < \infty \quad \text{and} \quad \int_1^\infty d\mu(s) < \infty. \tag{7}$$

Suppose first that μ is a positive measure on $]0,\infty[$ satisfying (5). For all $x, s>0$ we have

$$1-e^{-xs} \leqq xs \quad \text{and} \quad 1-e^{-xs} \leqq 2,$$

so by (7) the integral

$$g(x)=\int_0^\infty (1-e^{-xs})\,d\mu(s)$$

is finite for $x>0$. For $x>0$ and $h>0$ we have

$$\frac{1}{h}(g(x+h)-g(x))=\int_0^\infty \frac{1}{h}e^{-xs}(1-e^{-hs})\,d\mu(s),$$

and by the dominated convergence theorem we find

$$\lim_{h\to 0}\frac{1}{h}(g(x+h)-g(x))=\int_0^\infty s e^{-xs}\,d\mu(s). \tag{8}$$

A similar argument shows that (8) is also valid for $h<0$, so g is differentiable and

$$Dg(x)=\int_0^\infty s e^{-xs}\,d\mu(s) \quad \text{for } x>0.$$

It follows by Theorem 9.3 that Dg is completely monotone and g is thus a Bernstein function. The function

$$f(x)=a+bx+g(x),$$

where $a, b\geqq 0$, is consequently a Bernstein function.

Suppose next that f is a Bernstein function. Then Df is completely monotone, and by Theorem 9.3 there exists a positive measure ν on $[0,\infty[$ such that

$$Df(x)=\int_0^\infty e^{-xs}\,d\nu(s) \quad \text{for } x>0.$$

The measure ν can be written

$$\nu=b\varepsilon_0+\omega,$$

where $b=\nu(\{0\})\geqq 0$ and ω is the restriction of ν to $]0,\infty[$. We then have

$$Df(x)=b+\int_0^\infty e^{-xs}\,d\omega(s),$$

and putting $a=\lim_{x\to 0} f(x)$, we find for $x>0$

$$f(x)=a+\int_0^x Df(u)\,du=a+bx+\int_0^\infty \frac{1-e^{-xs}}{s}\,d\omega(s).$$

In particular we have

$$(1-e^{-1})\int_1^\infty \frac{1}{s}\,d\omega(s) \leqq \int_1^\infty \frac{1-e^{-s}}{s}\,d\omega(s) \leqq f(1),$$

which shows that

$$\int_1^\infty \frac{1}{s}\,d\omega(s) < \infty.$$

The measure μ on $]0,\infty[$ defined by $\mu = \frac{1}{s}\,d\omega(s)$ then clearly satisfies (7), and hence (5), and this shows that f has a representation of the form (6).

In the representation (6) of a Bernstein function f, the constants a and b are uniquely determined as

$$a = \lim_{x\to 0} f(x) \quad \text{and} \quad b = \lim_{x\to\infty} \frac{f(x)}{x},$$

and since

$$Df(x) = b + \int_0^\infty s e^{-xs}\,d\mu(s),$$

the measure $b\varepsilon_0 + s\,d\mu(s)$ is uniquely determined as the representing measure in Theorem 9.3 for the completely monotone function Df. It follows that μ is uniquely determined by f. □

9.9. Exercise. Let f be a non-zero Bernstein function. Then $f(x) > 0$ for all $x > 0$ and $1/f$ is completely monotone.

9.10. Exercise. Let f be a non-zero Bernstein function and let g be a completely monotone function. Then $g \circ f$ is completely monotone. For a converse to this result see Bochner [1] p. 84.

9.11. For a measure μ on $[0,\infty[$, such that the function $s \mapsto e^{-xs}$ is integrable with respect to $|\mu|$ for all $x > 0$, the function

$$x \mapsto \int_0^\infty e^{-xs}\,d\mu(s)$$

is called the *Laplace transform* of μ, and it is denoted $\mathscr{L}\mu$, i.e.

$$\mathscr{L}\mu(x) = \int_0^\infty e^{-xs}\,d\mu(s) \qquad \text{for } x > 0. \tag{9}$$

Since $|e^{-zs}| = e^{-(\operatorname{Re} z)s}$ for $s \geqq 0$, it is clear that the Laplace transform $\mathscr{L}\mu$ extends to a holomorphic function in the open half-plane $\operatorname{Re} z > 0$, given by the formula

$$\mathscr{L}\mu(z) = \int_0^\infty e^{-zs}\,d\mu(s). \tag{10}$$

The Laplace transform $\mathscr{L}\mu$ of a bounded measure μ extends to a continuous function, defined by the integral in (10), in the closed half-plane $\operatorname{Re} z \geqq 0$.

The function defined by (10) in the open half-plane $\operatorname{Re} z > 0$, and when μ is bounded in the closed half-plane $\operatorname{Re} z \geqq 0$, will be called the *canonical extension* of the function (9).

By (10) we get

$$\mathscr{L}\mu(x+iy) = \mathscr{F}_{\mathbb{R}}(e^{-xs}d\mu(s))(y) \tag{11}$$

for $x>0$ and $y \in \mathbb{R}$, and if μ is bounded (11) also holds for $x=0$. This equation relates the Laplace transformation $\mathscr{L}$ to the Fourier transformation $\mathscr{F}_{\mathbb{R}}$ and makes it possible to deduce properties of the Laplace transformation from properties of the Fourier transformation.

The set of measures μ on $[0, \infty[$ for which

$$\int_0^\infty e^{-xs} d|\mu|(s) < \infty \qquad \text{for all } x>0,$$

is an algebra under addition and convolution, and the Laplace transformation $\mathscr{L}$ is a homomorphism of this algebra into the algebra of holomorphic functions on the open half-plane $\operatorname{Re} z > 0$.

The Laplace transformation $\mathscr{L}$ is injective in the sense that if $\mathscr{L}\mu(x)=0$ for all $x>0$ then $\mu=0$.

We can now formulate the following corollaries to Theorem 9.3.

9.12. Corollary. *Every completely monotone function f has a canonical holomorphic extension to the open half-plane* $\operatorname{Re} z > 0$, *and f has a canonical continuous extension to the closed half-plane* $\operatorname{Re} z \geqq 0$ *if and only if* $\lim_{x\to 0} f(x) < \infty$.

9.13. Corollary. *The Laplace transformation $\mathscr{L}$ defines a bijection of the set of positive measure μ on $[0, \infty[$ for which*

$$\int_0^\infty e^{-xs} d\mu(s) < \infty \qquad \textit{for all } x>0$$

onto the set of completely monotone functions, and $\mathscr{L}$ defines a bijection of the set of positive bounded measures on $[0,\infty[$ onto the set of completely monotone functions f for which $\lim_{x\to 0} f(x) < \infty$.

9.14. Let $f\colon]0,\infty[\to \mathbb{R}$ be a Bernstein function. Then f has an integral representation of the form (6), which may be used to define a holomorphic extension of f.

For $s>0$ and $z \in \mathbb{C}$ with $\operatorname{Re} z \geqq 0$ we have the elementary inequalities

$$|1-e^{-zs}| \leqq s|z| \quad \text{and} \quad |1-e^{-zs}| \leqq 2,$$

which show that the right-hand side of (6) is well-defined in the half-plane $\operatorname{Re} z \geqq 0$.

The function defined in this half-plane by

$$z \mapsto a + bz + \int_0^\infty (1 - e^{-zs})\, d\mu(s) \tag{12}$$

is easily seen to be continuous, and to be holomorphic in the open half-plane $\operatorname{Re} z > 0$.

It follows that every Bernstein function has an extension to a continuous function on the closed half-plane $\operatorname{Re} z \geqq 0$, which is holomorphic in the open half-plane $\operatorname{Re} z > 0$. An extension with these properties is clearly unique and is called the *canonical extension.*

9.15. Remark. In the following we will use the canonical extensions of completely monotone functions and Bernstein functions without further mentioning. In particular we will use the same symbol for such a function on the interval $]0, \infty[$ and for the canonical extension.

9.16. Remark. Corollary 9.13 and Bochner's theorem in connection with (11) explain the analogy between the set of completely monotone functions and the set of continuous positive definite functions. The following result and Theorem 8.3 explain the analogy between Bernstein functions and continuous negative definite functions.

9.17. Definition. A convolution semigroup $(\eta_t)_{t>0}$ on $\mathbb{R}$ is said to be *supported by* $[0, \infty[$, if $\operatorname{supp}(\eta_t) \subseteq [0, \infty[$ for all $t > 0$.

9.18. Theorem. *There is a one-to-one correspondence between convolution semigroups* $(\eta_t)_{t>0}$ *on* $\mathbb{R}$ *supported by* $[0, \infty[$ *and Bernstein functions* $f\colon]0, \infty[\to \mathbb{R}$. *The correspondence is given by the formula*

$$\mathscr{L}\eta_t(x) = e^{-tf(x)} \quad \textit{for } t > 0 \text{ and } x > 0. \tag{13}$$

Proof. Suppose first that $(\eta_t)_{t>0}$ is a convolution semigroup on $\mathbb{R}$ supported by $[0, \infty[$, and consider for a fixed $x > 0$ the function $\varphi_x\colon]0, \infty[\to \mathbb{R}$ defined by

$$\varphi_x(t) = \mathscr{L}\eta_t(x) \quad \text{for } t > 0.$$

Then $\varphi_x(t) > 0$ for all $t > 0$, and furthermore

$$\varphi_x(s+t) = \varphi_x(s)\varphi_x(t) \quad \text{for } s, t > 0.$$

By Proposition 8.2 φ_x is continuous and $\lim_{t \to 0} \varphi_x(t) = 1$, and there exists consequently a uniquely determined real number $f(x)$ such that

$$\varphi_x(t) = \mathscr{L}\eta_t(x) = \exp(-tf(x)) \quad \text{for all } t > 0.$$

Since $\varphi_x(t) \leqq 1$ for all $x, t > 0$, we have that $f(x) \geqq 0$ for all $x > 0$, and it follows by Proposition 9.2, that f is a Bernstein function.

Suppose next that f is a Bernstein function. By Proposition 9.2 the function $x \mapsto \exp(-tf(x))$ is completely monotone for all $t > 0$ and

$$\lim_{x \to 0} \exp(-tf(x)) \leqq 1,$$

since $\lim_{x \to 0} f(x) \geqq 0$.

There exists consequently (cf. 9.3 and 9.4) a uniquely determined positive measure η_t on $[0, \infty[$ with $\eta_t([0, \infty[) \leqq 1$ such that

$$\mathscr{L}\eta_t(x) = \exp(-tf(x)) \quad \text{for } x > 0 \text{ and } t > 0.$$

By unicity of holomorphic continuations this formula is valid for x in the open half-plane $\operatorname{Re} z > 0$, and by continuity for x in the closed half-plane $\operatorname{Re} z \geqq 0$. By (11) we find

$$\mathscr{F}_{\mathbb{R}}\eta_t(y) = \mathscr{L}\eta_t(iy) = \exp(-tf(iy)) \quad \text{for } y \in \mathbb{R}, \tag{14}$$

which by Schoenberg's theorem (7.8) shows that the function $y \mapsto f(iy)$ is negative definite. Since it is clearly continuous, we get by Theorem 8.3 that $(\eta_t)_{t>0}$ is a convolution semigroup. □

9.19. Remark. The formula (13) is valid for all x in the closed half-plane $\operatorname{Re} z \geqq 0$. In particular we have

$$\eta_t([0, \infty[) = \mathscr{L}\eta_t(0) = \exp(-tf(0)) \quad \text{for } t > 0,$$

so that the measures η_t are probability measures if and only if $f(0) = 0$. The continuous negative definite function associated with $(\eta_t)_{t>0}$ is $y \mapsto f(iy)$ for $y \in \mathbb{R}$.

9.20. Let $(\mu_t)_{t>0}$ be a convolution semigroup on a LCA-group G with associated negative definite function $\psi: \Gamma \to \mathbb{C}$, and let $(\eta_t)_{t>0}$ be a convolution semigroup on $\mathbb{R}$ supported by $[0, \infty[$ with corresponding Bernstein function f. Since $\operatorname{Re}\psi(\gamma) \geqq 0$ for all $\gamma \in \Gamma$ the composite function $f \circ \psi$ exists and by (12) we have

$$f(\psi(\gamma)) = a + b\psi(\gamma) + \int_0^\infty (1 - e^{-\psi(\gamma)s})\, d\mu(s) \quad \text{for } \gamma \in \Gamma. \tag{15}$$

This formula, together with 7.7 and 7.8, shows that $f \circ \psi$ is a continuous negative definite function. The corresponding convolution semigroup on G is denoted $(\mu_t^f)_{t>0}$ and it is called the *semigroup subordinated to* $(\mu_t)_{t>0}$ *by means of* $(\eta_t)_{t>0}$. It can be expressed in terms of $(\mu_t)_{t>0}$ and $(\eta_t)_{t>0}$ in the following way:

9.21. Proposition. *The convolution semigroup $(\mu_t^f)_{t>0}$ subordinated to $(\mu_t)_{t>0}$ by means of $(\eta_t)_{t>0}$ is given by the vague integral*

$$\mu_t^f=\int_0^\infty \mu_s d\eta_t(s) \quad \text{for } t>0, \tag{16}$$

and the resolvent $(\rho_\lambda^f)_{\lambda>0}$ for $(\mu_t^f)_{t>0}$ is given by the vague integral

$$\rho_\lambda^f=\int_0^\infty \mu_s d\tau_\lambda(s) \quad \text{for } \lambda>0, \tag{17}$$

where $(\tau_\lambda)_{\lambda>0}$ is the resolvent for $(\eta_t)_{t>0}$.

Proof. Let $t>0$. The mapping

$$g\mapsto \int_0^\infty \langle \mu_s, g\rangle d\eta_t(s)$$

is a positive linear form on $C_c(G)$, i.e. a positive measure ν_t on G such that

$$\langle \nu_t, g\rangle=\int_0^\infty \langle \mu_s, g\rangle d\eta_t(s) \quad \text{for } g\in C_c(G).$$

Clearly $\nu_t(G)\leqq 1$, and the Fourier transform of ν_t is for $\gamma\in\Gamma$ given by

$$\begin{aligned}\hat\nu_t(\gamma)&=\int_0^\infty \hat\mu_s(\gamma)\,d\eta_t(s)=\int_0^\infty e^{-s\psi(\gamma)}d\eta_t(s)\\&=\mathscr{L}\eta_t(\psi(\gamma))=\exp(-t f(\psi(\gamma))),\end{aligned}$$

and it follows that $\nu_t=\mu_t^f$.

For $\lambda>0$ and $g\in C_c(G)$ we then find

$$\begin{aligned}\langle \rho_\lambda^f, g\rangle&=\int_0^\infty e^{-\lambda t}\langle \mu_t^f, g\rangle dt=\int_0^\infty \left(e^{-\lambda t}\int_0^\infty \langle \mu_s, g\rangle d\eta_t(s)\right)dt\\&=\int_0^\infty \langle \mu_s, g\rangle d\tau_\lambda(s),\end{aligned}$$

which shows the last assertion. □

9.22. Remark. As was seen in 9.20, the composite function $f\circ\psi$ of a Bernstein function f and a continuous negative definite function ψ on Γ is well-defined and it is a continuous negative definite function on Γ. This may be expressed by saying that the Bernstein functions *operate* on the set $CN(\Gamma)$.

Conversely it can be shown that the Bernstein functions are the only functions which operate on $CN(\Gamma)$. Or more precisely: Let f be a function defined in the closed half-plane $\operatorname{Re} z\geqq 0$ with the property that for every LCA-group Γ and for every $\psi\in CN(\Gamma)$ the composite function $f\circ\psi$ belongs to $CN(\Gamma)$. Then f is the canonical extension of a Bernstein function. Cf. Harzallah [1].

9.23. Examples. Here are some examples of Bernstein functions and the corresponding convolution semigroups.

1) The constant function $f(x)=a$, $a \geqq 0$, is a Bernstein function and the corresponding convolution semigroup is $(e^{-at}\varepsilon_0)_{t>0}$.

2) The function $f(x)=bx$, $b \geqq 0$, is a Bernstein function and the corresponding convolution semigroup is $(\varepsilon_{bt})_{t>0}$.

3) The function $f(x)=1-e^{-xs}$, $s \geqq 0$, is a Bernstein function. The corresponding convolution semigroup is called the *Poisson semigroup with jump s* and is given by

$$\eta_t = \sum_{k=0}^{\infty} e^{-t} \frac{t^k}{k!} \varepsilon_{sk} \quad \text{for } t>0.$$

4) The function $f(x)=\log(1+x)$ is a Bernstein function. This may be seen directly, but follows also from the elementary formula

$$\log(1+x) = \int_0^\infty (1-e^{-xs}) s^{-1} e^{-s} ds \quad \text{for } x>0.$$

The corresponding convolution semigroup $(\eta_t)_{t>0}$ is called the *Γ-semigroup.* For each $t>0$ the measure η_t has a density g_t with respect to Lebesgue measure ds given by

$$g_t(s) = \begin{cases} \dfrac{1}{\Gamma(t)} s^{t-1} e^{-s} & \text{for } s>0, \\ 0 & \text{for } s \leqq 0. \end{cases}$$

5) The function $f(x)=x^\alpha$ is a Bernstein function for $\alpha \in [0,1]$. This may be seen directly, but follows also for $\alpha \in]0,1[$ from the elementary formula

$$x^\alpha = \frac{\alpha}{\Gamma(1-\alpha)} \int_0^\infty (1-e^{-xs}) s^{-\alpha-1} ds \quad \text{for } x>0.$$

The corresponding convolution semigroup is called the *one-sided stable semigroup of order* α, and it is denoted $(\sigma_t^\alpha)_{t>0}$. For $t>0$ we have $\sigma_t^0 = e^{-t}\varepsilon_0$ and $\sigma_t^1 = \varepsilon_t$. For $\alpha = \frac{1}{2}$ the measure σ_t^α has a density h_t with respect to Lebesgue measure ds given by

$$h_t(s) = \begin{cases} \dfrac{1}{\sqrt{4\pi}} t s^{-\frac{3}{2}} \exp\left(-\dfrac{t^2}{4s}\right) & \text{for } s>0, \\ 0 & \text{for } s \leqq 0, \end{cases}$$

cf. Erdélyi *et al.* [1] p. 245.

For a convolution semigroup $(\mu_t)_{t>0}$ on G with associated negative definite function ψ, the convolution semigroup associated with ψ^α, for $\alpha \in [0,1]$, is given by (cf. (16))

$$\int_0^\infty \mu_s \, d\sigma_t^\alpha(s) \quad \text{for } t>0. \tag{18}$$

9.24. Exercise. Let f be a completely monotone function, and suppose that $\lim_{x \to 0} f(x) \leqq a < \infty$. Then $a - f$ is a bounded Bernstein function. Conversely, every bounded Bernstein function can be written in this way. Every Bernstein function is the pointwise limit of Bernstein function of the form $a - f$, where f is completely monotone such that $\lim_{x \to 0} f(x) \leqq a$.

9.25. Exercise. Let $(\eta_t')_{t>0}$ and $(\eta_t'')_{t>0}$ be convolution semigroups supported by $[0, \infty[$ and let f_1 and f_2 be the corresponding Bernstein functions. Then $f_2 \circ f_1$ is a Bernstein function and the corresponding convolution semigroup $(\tau_t)_{t>0}$ is given by the vague integral

$$\tau_t = \int_0^\infty \eta_s' \, d\eta_t''(s) \quad \text{for } t > 0.$$

9.26. Further information about the Laplace transformation and completely monotone functions can be found in the book of Widder [1]. The (non-zero) Bernstein functions are called completely monotone mappings in Bochner [1], where Theorem 9.8 is proved. The notion of a convolution semigroup subordinated to another convolution semigroup goes back to Bochner, cf. Feller [1] p. 335.

§ 10. Examples of Negative Definite Functions and Convolution Semigroups

The present paragraph is mainly devoted to concrete examples of continuous, negative definite functions and their corresponding convolution semigroups on the LCA-group $\mathbb{R}^n$ ($n \geqq 1$). We shall also give a "geometric" criterion for negative definiteness, analogous to Polya's criterion for positive definiteness, and the paragraph is finished by mentioning the classical Lévy-Khinchin representation of continuous, negative definite functions on $\mathbb{R}^n$.

10.1. Let G be a LCA-group with dual group Γ, and consider a family $(\mu_t)_{t>0}$ of positive bounded measures on G. Using the Fourier transformation, it is often easy to decide whether $(\mu_t)_{t>0}$ is a convolution semigroup or not. Condition (ii) of Definition 8.1 is equivalent with

$$\hat{\mu}_t \cdot \hat{\mu}_s = \hat{\mu}_{t+s} \quad \text{for } t, s > 0,$$

because the Fourier transformation is injective, while condition (iii) of 8.1 is equivalent with

$$\lim_{t \to 0} \hat{\mu}_t(\gamma) = 1 \quad \text{for } \gamma \in \Gamma,$$

on account of the continuity theorem of Lévy (3.14).

10.2. The following table contains some important convolution semigroups $(\mu_t)_{t>0}$ on $\mathbb{R}$ and their associated continuous, negative definite functions ψ on $\mathbb{R}(\approx \hat{\mathbb{R}})$, cf. 9.23 and 10.5 below. The independent variable in ψ is denoted y.

Table 2

Convolution semigroup	$(\mu_t)_{t>0}$	ψ
Degenerate semigroup	$\mu_t=e^{-at}\varepsilon_0,\quad a\geqq 0$	a
Translation semigroup with speed $a\in\mathbb{R}$	$\mu_t=\varepsilon_{at}$	iay
Poisson's semigroup with jump $s\geqq 0$	$\mu_t=\sum_{k=0}^{\infty} e^{-t}\frac{t^k}{k!}\varepsilon_{sk}$	$1-e^{-isy}$
One-sided stable semigroup of order $\alpha\in[0,1]$	$\mu_t=\sigma_t^\alpha$	$(iy)^\alpha$
The Γ-semigroup	$\mu_t=1_{]0,\infty[}(x)\frac{1}{\Gamma(t)}x^{t-1}e^{-x}\,dx$	$\log(1+y^2)+i\operatorname{Arctan} y$
Brownian semigroup	$\mu_t=(4\pi t)^{-\frac{1}{2}}\exp\left(-\frac{x^2}{4t}\right)dx$	y^2
Symmetric stable semigroup of order $\alpha\in]0,2[$	$\mu_t=\mu_t^\alpha$	$\|y\|^\alpha$
Cauchy's semigroup	$\mu_t=\frac{t}{\pi}(t^2+x^2)^{-1}\,dx$	$\|y\|$

In the following sections we shall see how these "one-dimensional" convolution semigroups can be combined or transformed into convolution semigroups on $\mathbb{R}^n$. The euclidean norm and inner product in $\mathbb{R}^n$ is denoted $\|\cdot\|$ and $(\cdot,\cdot)$ respectively.

10.3. Consider the family $(\mu_t)_{t>0}$ of measures on $\mathbb{R}^n$ ($n\geqq 1$) having the functions $(g_t)_{t>0}$ defined by

$$g_t(x)=(4\pi t)^{-\frac{n}{2}}\exp\left(-\frac{\|x\|^2}{4t}\right)\quad\text{for } x\in\mathbb{R}^n,$$

as densities with respect to the Lebesgue measure on $\mathbb{R}^n$.

For the Fourier transform of g_t we find

$$\begin{aligned}\hat{g}_t(y)&=\int_{\mathbb{R}^n}\exp(-i(x,y))(4\pi t)^{-\frac{n}{2}}\exp\left(-\frac{\|x\|^2}{4t}\right)dx\\&=\prod_{k=1}^{n}\int_{\mathbb{R}}\exp(-ix_k y_k)(4\pi t)^{-\frac{1}{2}}\exp\left(-\frac{x_k^2}{4t}\right)dx_k\\&=\prod_{k=1}^{n}\exp(-t y_k^2)=\exp(-t\|y\|^2).\end{aligned}$$

It follows that $(\mu_t)_{t>0}$ is a convolution semigroup on $\mathbb{R}^n$, the *Brownian* (or *Gaussian*) semigroup on $\mathbb{R}^n$. The associated continuous, negative definite function $\psi\colon\mathbb{R}^n\to\mathbb{C}$ is the function $y\mapsto\|y\|^2$.

10.4. The family $(\mu_t')_{t>0}$ of measures on $\mathbb{R}^{n+1}$ $(n \geq 1)$ defined by

$$\mu_t' = \mu_t \otimes \varepsilon_t \quad \text{for } t>0,$$

where $(\mu_t)_{t>0}$ is the Brownian semigroup on $\mathbb{R}^n$, is clearly a convolution semigroup on $\mathbb{R}^{n+1}$, the *heat semigroup* on $\mathbb{R}^{n+1}$. The associated continuous, negative definite function on $\mathbb{R}^{n+1}$ is the function

$$(y,s) \mapsto \|y\|^2 + is \quad \text{for } (y,s) \in \mathbb{R}^n \times \mathbb{R},$$

cf. Exercise 8.9.

10.5. By 9.20 and 9.23 the function $y \mapsto \|y\|^\alpha$ is continuous and negative definite on $\mathbb{R}^n$ for all $\alpha \in]0,2]$. The associated convolution semigroup $(\mu_t^\alpha)_{t>0}$ on $\mathbb{R}^n$ is called the *symmetric stable semigroup of order* α, and it is characterized by

$$\hat{\mu}_t^\alpha(y) = \exp(-t\|y\|^\alpha) \quad \text{for } t>0 \text{ and } y \in \mathbb{R}^n.$$

The symmetric stable semigroup of order α is the convolution semigroup subordinated to the Brownian semigroup $(\mu_t)_{t>0}$ by means of the one-sided stable semigroup $(\sigma_t^{\alpha/2})_{t>0}$ of order $\alpha/2$, cf. 9.23, and we therefore have

$$\mu_t^\alpha = \int_0^\infty \mu_s \, d\sigma_t^{\alpha/2}(s) \quad \text{for } t>0 \text{ and } \alpha \in]0,2], \tag{1}$$

cf. (18) of paragraph 9. In particular, the symmetric stable semigroup of order 2 is the Brownian semigroup.

It follows from Schoenberg's theorem (7.8) that the function $x \mapsto \exp(-\|x\|^\alpha)$ on $\mathbb{R}^n$ is positive definite for $\alpha \in]0,2]$. Cf. 5.5.

Since the function $\hat{\mu}_t^\alpha$ is integrable we get by Theorem 2.6 that

$$\mu_t^\alpha = g_t^\alpha(x)\, dx,$$

where

$$g_t^\alpha(x) = \frac{1}{(2\pi)^n} \int_{\mathbb{R}^n} \exp(i(x,y)) \exp(-t\|y\|^\alpha)\, dy$$

is a function belonging to $C_0^+(\mathbb{R}^n)$.

In the case $\alpha = 1$ we find by (1) using the expression for $\sigma_t^{\frac{1}{2}}$ that

$$\begin{aligned} g_t^1(x) &= t(4\pi)^{-\frac{n+1}{2}} \int_0^\infty s^{-\frac{n+3}{2}} \exp\left(-\frac{\|x\|^2+t^2}{4s}\right) ds \\ &= \Gamma\left(\frac{n+1}{2}\right) t[\pi(\|x\|^2+t^2)]^{-\frac{n+1}{2}}, \end{aligned}$$

which for $n=1$ reduces to the density for the Cauchy semigroup.

For other $\alpha \in]0,2[$ explicit expressions for the densities g_t^α are not known, but g_t^α can be expressed as sum of infinite series. Cf. Feller [1] p. 548.

10.6. Proposition. *A continuous function* $\psi: \mathbb{R} \to [0, \infty[$ *with the properties*

(i) ψ *is even, i.e.* $\psi(y) = \psi(-y)$ *for* $y \in \mathbb{R}$,

(ii) ψ *is increasing and concave on* $[0, \infty[$,

is negative definite.

Proof. For all $n \in \mathbb{N}$ the function $\psi_n = \inf(\psi, n)$ is continuous and even on $\mathbb{R}$, and increasing and concave on $[0, \infty[$. The function $\varphi_n = n - \psi_n$ is therefore continuous, even and non-negative on $\mathbb{R}$ and decreasing and convex on $[0, \infty[$. By Polya's theorem (cf. 5.4) it follows that φ_n is positive definite, and Corollary 7.7 then gives that

$$\psi_n = \psi_n - \psi_n(0) + \psi_n(0) = \varphi_n(0) - \varphi_n + \psi_n(0)$$

is negative definite. We conclude by Proposition 7.4 because

$$\psi(y) = \lim_{n \to \infty} \psi_n(y) \quad \text{pointwise on } \mathbb{R}. \quad \square$$

10.7. Exercise. The function $\psi: \mathbb{R} \to \mathbb{R}$ defined by

$$\psi(y) = \begin{cases} |y| & \text{if } |y| \leqq 1, \\ 1 & \text{if } |y| > 1, \end{cases}$$

is continuous and negative definite. Find the associated convolution semigroup on $\mathbb{R}$.

The set of all continuous, negative definite functions on $\mathbb{R}^n$ is described by the so-called *Lévy-Khinchin* formula.

10.8. Theorem. *Let* $\psi: \mathbb{R}^n \to \mathbb{C}$ *be a continuous, negative definite function. There exist*

(i) *a constant* $c \geqq 0$,

(ii) *a continuous linear form* $l: \mathbb{R}^n \to \mathbb{R}$

$$l(y) = \sum_{k=1}^{n} b_k y_k \quad \text{with } b_k \in \mathbb{R},$$

(iii) *a continuous, non-negative quadratic form* $q: \mathbb{R}^n \to \mathbb{R}$

$$q(y) = \sum_{j,k=1}^{n} a_{jk} y_j y_k \quad \text{with } a_{jk} \in \mathbb{R} \text{ and } a_{jk} = a_{kj},$$

(iv) *a non-negative, bounded measure* μ *on* $\mathbb{R}^n \setminus \{0\}$,

such that for $y \in \mathbb{R}^n$

$$\psi(y) = c + i l(y) + q(y) + \int_{\mathbb{R}^n \setminus \{0\}} \left[1 - \exp(-i(x, y)) - \frac{i(x, y)}{1 + \|x\|^2}\right] \frac{1 + \|x\|^2}{\|x\|^2} d\mu(x). \quad (2)$$

Conversely, if (c, l, q, μ) is a quadruple as specified above, then (2) defines a continuous, negative definite function. Furthermore the quadruple (c, l, q, μ) is uniquely determined by ψ.

For the proof we refer to Courrège [1].

10.9. Remark. By Theorem 9.8 and Remark 9.19 we get the following formula for the set of continuous negative definite functions ψ on $\mathbb{R}$ associated with convolution semigroups on $\mathbb{R}$ supported by $[0, \infty[$:

$$\psi(x) = a + ibx + \int_0^\infty (1 - e^{-ixs})\, d\mu(s) \quad \text{for } x \in \mathbb{R}, \tag{3}$$

where $a, b \geqq 0$ and μ is a positive measure on $]0, \infty[$ satisfying

$$\int_0^\infty \frac{s}{1+s}\, d\mu(s) < \infty.$$

Formula (3) is a special case of (2).

10.10. The Lévy-Khinchin formula (2) in the case $n=1$ was established by P. Lévy in 1934 and A. J. Khinchin in 1937. It has been extended to Lie groups by Hunt [1] and by Parthasarathy *et al.* [1] to locally compact abelian groups with a countable case. See also the book of Parthasarathy [1]. We shall later give a representation formula, due to Harzallah [2] for the real continuous negative definite functions on an arbitrary locally compact abelian group, cf. 18.20. Recently Hazod [1] has obtained a Lévy-Khinchin formula for an arbitrary locally compact group.

§ 11. Contraction Semigroups

Notions and results from the theory of semigroups of operators will be indispensable for some of the following paragraphs, and we shall therefore give a short introduction to this theory.

Let E be a fixed Banach space over the complex numbers with norm $\|\cdot\|$. The reader is assumed to be familiar with integration of continuous E-valued functions, cf. e.g. Bourbaki [2].

For a given semigroup $(P_t)_{t>0}$ on E we introduce the infinitesimal generator $(A, D(A))$ and the resolvent $(N_\lambda)_{\lambda>0}$ and furthermore two operators, the potential operator $(N, D(N))$ and the zero-resolvent $(N_0, D(N_0))$, which both play the role of an "inverse" operator to $-A$.

11.1. Definition. A *contraction semigroup* on E is a family $(P_t)_{t>0}$ of bounded operators on E satisfying

(i) $\|P_t\| \leqq 1$ for $t>0$,

(ii) $P_t P_s = P_{t+s}$ for $t, s>0$.

The contraction semigroup $(P_t)_{t>0}$ on E is said to be *strongly continuous* if

(iii) $\lim_{t\to 0} P_t f = f$ for all $f \in E$.

It is clear that the operators in a contraction semigroup are mutually commuting.

11.2. Lemma. *Let $(P_t)_{t>0}$ be a contraction semigroup on E. The set*

$$E_0 = \{f \in E \mid \lim_{t \to 0} P_t f = f\}$$

is a closed subspace of E, and putting $P_0 = I$ (the identity operator) the mapping $t \mapsto P_t f$ of $[0, \infty[$ into E is continuous for all $f \in E_0$.

Proof. It is clear that E_0 is a subspace of E. For a given $f \in \bar{E}_0$ and $\varepsilon > 0$ there exist $g \in E_0$ such that $\|f - g\| < \varepsilon$ and $t_0 > 0$ with the property that $\|P_t g - g\| < \varepsilon$ for $t \in [0, t_0]$. Since $\|P_t\| \leqq 1$ we find for all $t \in [0, t_0]$

$$\|P_t f - f\| \leqq \|P_t f - P_t g\| + \|P_t g - g\| + \|g - f\| < 3\varepsilon,$$

which shows that $f \in E_0$, i.e. E_0 is closed.

The inequality

$$\|P_t f - P_{t_0} f\| = \|P_{t_0}(P_{t-t_0} f - f)\| \leqq \|P_{t-t_0} f - f\|$$

for $0 \leqq t_0 \leqq t$ and the analogous inequality for $0 \leqq t \leqq t_0$ show that the mapping $t \mapsto P_t f$ is continuous for $f \in E_0$. □

11.3. Remark. For a strongly continuous contraction semigroup $(P_t)_{t>0}$ on E we put $P_0 = I$, and the mapping $t \mapsto P_t$ of $[0, \infty[$ into the set of bounded linear transformations of E is strongly continuous, as shown by Lemma 11.2.

In order to verify that a given contraction semigroup $(P_t)_{t>0}$ on E is strongly continuous, it is enough, by the above properties of E_0, to prove that $\lim_{t \to 0} P_t f = f$ for all f belonging to a total subset of E.

We shall now introduce some important operators associated with a fixed strongly continuous contraction semigroup $(P_t)_{t>0}$ on E. An operator on E is written $(B, D(B))$. Here $D(B)$ is the domain of the operator and the range of $(B, D(B))$ is denoted $R(B)$, i.e. $R(B) = B(D(B))$.

11.4. Definition. The *infinitesimal generator* $(A, D(A))$ for $(P_t)_{t>0}$ is the operator on E with domain

$$D(A) = \left\{ f \in E \mid \lim_{t \to 0} \frac{1}{t}(P_t f - f) \text{ exists in } E \right\}$$

and given by

$$Af = \lim_{t \to 0} \frac{1}{t}(P_t f - f) \quad \text{for } f \in D(A).$$

11.5. Definition. The *potential operator* $(N, D(N))$ for $(P_t)_{t>0}$ is the operator on E with domain

$$D(N) = \left\{ f \in E \mid \lim_{t \to \infty} \int_0^t P_s f \, ds \text{ exists in } E \right\}$$

and given by

$$Nf = \lim_{t\to\infty} \int_0^t P_s f\,ds \qquad \text{for } f\in D(N).$$

It is clear that $D(A)$ and $D(N)$ are subspaces of E, and that $A: D(A)\to E$ and $N: D(N)\to E$ are linear mappings.

11.6. Proposition. *The infinitesimal generator $(A, D(A))$ is a densely defined and closed operator on E.*

Proof. Let $f\in E$ and consider for $a>0$ the vector

$$f^a = \int_0^a P_s f\,ds \in E.$$

An easy computation gives that

$$\frac{1}{t}(P_t f^a - f^a) = \frac{1}{t}\int_a^{a+t} P_s f\,ds - \frac{1}{t}\int_0^t P_s f\,ds \qquad \text{for } t>0,$$

and as the right-hand side converges to $P_a f - f$ for $t\to 0$ we have $f^a\in D(A)$ and $Af^a = P_a f - f$. Furthermore

$$\frac{1}{a} f^a \to f \qquad \text{for } a\to 0,$$

and $D(A)$ is consequently dense in E.

To prove that $(A, D(A))$ is a closed operator in E, we first remark that, as is easily seen, $P_t(D(A))\subseteq D(A)$ for $t>0$, and that the mapping $t\mapsto P_t f$ for $f\in D(A)$ is differentiable with the continuous derivative

$$\frac{d}{dt}(P_t f) = P_t(Af) = A(P_t f).$$

It follows that for $f\in D(A)$ and $t>0$

$$P_t f - f = \int_0^t A(P_s f)\,ds = \int_0^t P_s(Af)\,ds. \tag{1}$$

Consider now a sequence $(f_n)_{n\in\mathbb{N}}$ of elements from $D(A)$ with the properties that $(f_n)_{n\in\mathbb{N}}$ converges to some $f\in E$ and $(Af_n)_{n\in\mathbb{N}}$ converges to some $g\in E$. We shall show that $f\in D(A)$ and $Af=g$. By equation (1) we have for $t>0$

$$\frac{1}{t}(P_t f_n - f_n) = \frac{1}{t}\int_0^t P_s(Af_n)\,ds,$$

hence letting $n \to \infty$

$$\frac{1}{t}(P_t f - f) = \frac{1}{t}\int_0^t P_s g\, ds.$$

The right-hand side converges to g as $t \to 0$, and this shows that $f \in D(A)$ and $Af = g$. □

The situation is not so simple for the potential operator. We start by giving some elementary properties of $(N, D(N))$.

11.7. Lemma.
(i) *$P_t(D(N)) \subseteq D(N)$ for $t > 0$.*
(ii) *$P_t(Nf) = N(P_t f)$ for $t > 0$ and $f \in D(N)$.*
(iii) *$P_t(Nf) - Nf = -\int_0^t P_s f\, ds$ for $t > 0$ and $f \in D(N)$.*
(iv) *$D(N) \subseteq \overline{R(N)}$.*
(v) *$\lim_{t \to 0} P_t f = 0$ for $f \in \overline{R(N)}$.*
(vi) *$R(N) \subseteq D(A)$ and $A(Nf) = -f$ for $f \in D(N)$; in particular N is injective.*

Proof. Let $t > 0$ and $f \in D(N)$. For $a > 0$ we have

$$\int_0^a P_s(P_t f)\, ds = P_t\left(\int_0^a P_s f\, ds\right),$$

and since the right-hand side converges to $P_t(Nf)$ as $a \to \infty$, we find that $P_t f \in D(N)$ and $N(P_t f) = P_t(Nf)$, thus proving (i) and (ii).

For $f \in D(N)$ and $t > 0$ we get using (ii) that

$$\begin{aligned} P_t(Nf) &= \lim_{a \to \infty} \int_0^a P_{s+t} f\, ds = \lim_{a \to \infty} \int_t^{a+t} P_s f\, ds \\ &= \lim_{a \to \infty} \left(\int_0^{a+t} P_s f\, ds - \int_0^t P_s f\, ds \right) = Nf - \int_0^t P_s f\, ds, \end{aligned}$$

so by (ii)

$$N\left(\frac{1}{t}(f - P_t f)\right) = \frac{1}{t}\int_0^t P_s f\, ds,$$

and since the right-hand side converges to f as $t \to 0$, it follows that f is limit of elements from $R(N)$ i.e. $f \in \overline{R(N)}$.

The set $\{f \in E \mid \lim_{t \to \infty} P_t f = 0\}$ is easily seen to be closed, and by (iii) $\lim_{t \to \infty} P_t(Nf) = 0$ for $f \in D(N)$, hence

$$\overline{R(N)} \subseteq \{f \in E \mid \lim_{t \to \infty} P_t f = 0\}.$$

For $f \in D(N)$ we have for all $t > 0$ by (iii) that

$$\frac{1}{t}(P_t(Nf) - Nf) = -\frac{1}{t}\int_0^t P_s f\, ds,$$

and since the right-hand side converges to $-f$ for $t \to 0$ we find $Nf \in D(A)$ and $A(Nf) = -f$. □

11.8. Lemma. *For $f\in D(N)$ and $a>0$ we have*

$$f^a=\int_0^a P_s f\,ds\in D(N).$$

Proof. For $t>0$ we find

$$\int_0^t P_u f^a du=\int_0^t\left(\int_0^a P_{u+s} f\,ds\right)du=\int_0^a\left(\int_0^t P_{u+s} f\,du\right)ds$$
$$=\int_0^a\left(\int_s^{s+t} P_u f\,du\right)ds=\int_0^a (f^{s+t}-f^s)ds$$
$$=\int_0^a f^{s+t}ds-\int_0^a f^s ds.$$

The proof will be finished when we have seen that $\int_0^a f^{s+t}ds$ converges to some element of E as $t\to\infty$. Let $\varepsilon>0$ be given and choose $t_0>0$ such that $\|Nf-f^t\|<\varepsilon$ for $t\geqq t_0$. For all $t\geqq t_0$ we then have

$$\left\|\int_0^a f^{s+t}ds-aNf\right\|=\left\|\int_t^{a+t}(f^s-Nf)ds\right\|\leqq a\varepsilon. \quad \square$$

11.9. Proposition. *The following three conditions are equivalent:*
(i) *$D(N)$ is dense in E.*
(ii) *$R(N)$ is dense in E.*
(iii) *$\lim_{t\to\infty} P_t f=0$ for all $f\in E$.*

When conditions (i)–(iii) are fulfilled the potential operator N is a densely defined, closed operator in E, and the infinitesimal generator A is injective and satisfies

$$N=-A^{-1} \quad \textit{and} \quad A=-N^{-1}.$$

Proof. The implications (i) ⇒ (ii) and (ii) ⇒ (iii) are clear by Lemma 11.7 (iv) and (v).

Suppose (iii). By equation (1) we have for $f\in D(A)$ and $t>0$

$$\int_0^t P_s(Af)ds=P_t f-f,$$

and it follows by (iii) that

$$\lim_{t\to\infty}\int_0^t P_s(Af)ds=-f, \tag{2}$$

i.e. $Af\in D(N)$ and $N(Af)=-f$. In particular A is an injective mapping of $D(A)$ into $D(N)$, and this together with Lemma 11.7 (vi) show that

$$N=-A^{-1} \quad \text{and} \quad A=-N^{-1}.$$

The closedness of N follows from the closedness of A. Equation (2) above shows that every $f \in D(A)$ is limit of vectors which by Lemma 11.8 belong to $D(N)$, i.e. $D(A) \subseteq \overline{D(N)}$, and it follows that $D(N)$ is dense in E (cf. Proposition 11.6). □

For a strongly continuous contraction semigroup $(P_t)_{t>0}$ on E with infinitesimal generator $(A, D(A))$ and potential operator $(N, D(N))$ we define for every $\lambda > 0$ a strongly continuous contraction semigroup $(P_t^\lambda)_{t>0}$ on E by

$$P_t^\lambda f = e^{-\lambda t} P_t f \quad \text{for } f \in E \text{ and } t > 0.$$

The infinitesimal generator, respectively the potential operator for $(P_t^\lambda)_{t>0}$ is denoted $(A_\lambda, D(A_\lambda))$ and $(N_\lambda, D(N_\lambda))$.

11.10. Proposition. *For all $\lambda > 0$ we have*

(i) *$D(A_\lambda) = D(A)$ and $A_\lambda = A - \lambda I$.*

(ii) *$D(N_\lambda) = E$ and N_λ is a bounded operator of norm $\|N_\lambda\| \leq \frac{1}{\lambda}$ on E which is given by*

$$N_\lambda f = \int_0^\infty e^{-\lambda t} P_t f \, dt \quad \text{for } f \in E.$$

(iii) *$\lim_{t \to \infty} P_t^\lambda f = 0$ for all $f \in E$.*

The proof is almost immediate.

11.11. The family $(N_\lambda)_{\lambda > 0}$ of bounded operators on E from Proposition 11.10 is called the *resolvent for* $(P_t)_{t>0}$. Using the fact that $N_\lambda = (\lambda I - A)^{-1}$ (by 11.10 (iii) we may apply Proposition 11.9) it is easy to see that

$$N_\lambda - N_\mu = (\mu - \lambda) N_\lambda N_\mu \quad \text{for } \lambda, \mu > 0, \tag{3}$$

and

$$N_\lambda N_\mu = N_\mu N_\lambda \quad \text{for } \lambda, \mu > 0. \tag{4}$$

Equation (3) is called the *resolvent equation.*

Furthermore we have

$$\lim_{\lambda \to \infty} \lambda N_\lambda f = f \quad \text{for all } f \in E, \tag{5}$$

which follows from the formula

$$\lambda N_\lambda f = \int_0^\infty \lambda e^{-\lambda t} P_t f \, dt = \int_0^\infty e^{-u} P_{u/\lambda} f \, du,$$

by use of the dominated convergence theorem.

11.12. Definition. The *zero-resolvent* $(N_0, D(N_0))$ for $(P_t)_{t>0}$ is the operator on E with domain

$$D(N_0)=\{f\in E \mid \lim_{\lambda\to 0} N_\lambda f \text{ exists in } E\}$$

and given by

$$N_0 f=\lim_{\lambda\to 0} N_\lambda f \quad \text{for } f\in D(N_0).$$

11.13. Lemma.
(i) $N_\lambda(D(N_0))\subseteq D(N_0)$ *for* $\lambda>0$.
(ii) $N_\lambda(N_0 f)=N_0(N_\lambda f)$ *for* $\lambda>0$ *and* $f\in D(N_0)$.
(iii) $N_0 f=N_\lambda f+\lambda N_\lambda(N_0 f)$ *for* $\lambda>0$ *and* $f\in D(N_0)$.
(iv) $D(N_0)\subseteq\overline{R(N_0)}$.
(v) $\lim_{\lambda\to 0}\lambda N_\lambda f=0$ *for* $f\in\overline{R(N_0)}$.
(vi) $R(N_0)\subseteq D(A)$ *and* $A(N_0 f)=-f$ *for* $f\in D(N_0)$; *in particular* N_0 *is injective.*

Proof. (i) and (ii) follow immediately from (4), and (iii) is a simple consequence of the resolvent equation (3).

By (ii) and (iii) we have for $f\in D(N_0)$ and $\lambda>0$

$$N_0(\lambda(f-\lambda N_\lambda f))=\lambda N_\lambda f,$$

and it follows by (5) that f is limit of elements from $R(N_0)$, i.e. $f\in\overline{R(N_0)}$.

The set $\{f\in E \mid \lim_{\lambda\to 0}\lambda N_\lambda f=0\}$ is easily seen to be closed, and by (iii) we get for $f\in D(N_0)$ that

$$\lim_{\lambda\to 0}\lambda N_\lambda(N_0 f)=N_0 f-\lim_{\lambda\to 0} N_\lambda f=0,$$

hence

$$\overline{R(N_0)}\subseteq\{f\in E \mid \lim_{\lambda\to 0}\lambda N_\lambda f=0\}.$$

From the formula $N_\lambda=(\lambda I-A)^{-1}$ it follows that

$$A(N_\lambda f)=-f+\lambda N_\lambda f \quad \text{for } f\in E.$$

For $f\in D(N_0)$ we therefore have

$$\lim_{\lambda\to 0} A(N_\lambda f)=-f,$$

and by the closedness of A (cf. 11.6), this implies that $N_0 f\in D(A)$ and $A(N_0 f)=-f$. □

11.14. Proposition. *The following three conditions are equivalent:*
(i) $D(N_0)$ *is dense in* E.
(ii) $R(N_0)$ *is dense in* E.
(iii) $\lim_{\lambda\to 0}\lambda N_\lambda f=0$ *for all* $f\in E$.

When the conditions (i)–(iii) *are fulfilled, the zero-resolvent* N_0 *is a densely defined closed operator in E, and the infinitesimal generator A is injective and satisfies*

$$N_0 = -A^{-1} \quad \text{and} \quad A = -N_0^{-1}.$$

Proof. The implications (i)⇒(ii) and (ii)⇒(iii) are clear by the preceding lemma.

Suppose (iii). For $f \in D(A)$ we have

$$A(N_\lambda f) = N_\lambda(Af) = -f + \lambda N_\lambda f \quad \text{for } \lambda > 0,$$

and it follows that $\lim_{\lambda \to 0} N_\lambda(Af) = -f$, i.e. $Af \in D(N_0)$ and $N_0(Af) = -f$. In particular A is an injective mapping of $D(A)$ into $D(N_0)$, and this together with Lemma 11.13 (vi) show that

$$N_0 = -A^{-1} \quad \text{and} \quad A = -N_0^{-1},$$

and it follows in particular that N_0 is a closed operator.

For $f \in D(A)$ we also have

$$\lim_{\lambda \to 0} A(N_\lambda f) = -f,$$

and since $A(N_\lambda f) \in D(N_0)$ we get that $D(A) \subseteq \overline{D(N_0)}$. The operator A being densely defined (cf. 11.6), we have that $\overline{D(N_0)} = E$. □

11.15. Proposition. *The zero-resolvent* $(N_0, D(N_0))$ *is an extension of the potential operator* $(N, D(N))$. *If* $\lim_{t \to \infty} P_t f = 0$ *for all* $f \in E$ *we have*

$$N = N_0 = -A^{-1}.$$

Proof. Let $f \in D(N)$. We shall prove that $f \in D(N_0)$ and $N_0 f = N f$. Putting

$$f^t = \int_0^t P_s f \, ds \quad \text{for } t > 0,$$

we have $\lim_{t \to \infty} f^t = N f$, and by partial integration we find for each $\lambda > 0$ and $t > 0$

$$\int_0^t e^{-\lambda s} P_s f \, ds = [e^{-\lambda s} f^s]_0^t + \int_0^t \lambda e^{-\lambda s} f^s \, ds.$$

For $t \to \infty$ we get

$$N_\lambda f = \int_0^\infty \lambda e^{-\lambda s} f^s \, ds = \int_0^\infty e^{-u} f^{u/\lambda} \, du,$$

and it follows that $\lim_{\lambda \to 0} N_\lambda f = N f$, the function $s \mapsto f^s$ being bounded.

The second part of the proposition follows from the Propositions 11.9 and 11.14 since the hypothesis $\lim_{t\to\infty} P_t f=0$ for $f\in E$ implies

$$\lim_{\lambda\to 0} \lambda N_\lambda f = \lim_{\lambda\to 0} \int_0^\infty e^{-u} P_{u/\lambda} f\, du = 0. \quad \square$$

11.16. Exercise. The operators $(N, D(N))$ and $(N_0, D(N_0))$ are closed also when they are not densely defined.

11.17. Exercise. We have $R(A)=R(\lambda N_\lambda - I)$ for all $\lambda>0$ and

$$\overline{R(A)}=\{f\in E \mid \lim_{\lambda\to 0} \lambda N_\lambda f=0\}.$$

We shall now indicate a characterization of the families $(N_\lambda)_{\lambda>0}$ of bounded operators on E, which are resolvents for strongly continuous contraction semigroups on E.

11.18. Definition. A *contraction resolvent* on E is a family $(N_\lambda)_{\lambda>0}$ of bounded operators on E satisfying

(i) $\|\lambda N_\lambda\| \leqq 1$ for $\lambda>0$.

(ii) $N_\lambda - N_\mu = (\mu-\lambda) N_\lambda N_\mu$ for $\lambda, \mu>0$ (*the resolvent equation*).

The contraction resolvent $(N_\lambda)_{\lambda>0}$ on E is called *strongly continuous* if

(iii) $\lim_{\lambda\to\infty} \lambda N_\lambda f = f$ for $f\in E$.

The resolvent for a strongly continuous contraction semigroup is a strongly continuous contraction resolvent.

By the so-called Hille-Yosida theorem, which we formulate below, the converse is also true. A special case of this theorem was encountered in 8.13.

11.19. Theorem. *Let $(N_\lambda)_{\lambda>0}$ be a strongly continuous contraction resolvent on E. There exists a uniquely determined strongly continuous contraction semigroup $(P_t)_{t>0}$ on E such that $(N_\lambda)_{\lambda>0}$ is the resolvent for $(P_t)_{t>0}$. The operator P_t is given by*

$$P_t f = \lim_{\lambda\to\infty} e^{-\lambda t} \exp(t\lambda^2 N_\lambda f) \quad \text{for } f\in E,$$

and the convergence is uniform for t in bounded intervals.

For the proof we refer to Meyer [1] or Yosida [2].

11.20. The above discussion of contraction semigroups covers only those aspects of the theory which will be essential in the sequel, and further information can be found in Hille and Phillips [1] or Yosida [2].

The potential operator has been studied in Berg [2] where Proposition 11.9 was proved, and the zero-resolvent was introduced and studied in Yosida [1].

§ 12. Translation Invariant Contraction Semigroups

A convolution semigroup $(\mu_t)_{t>0}$ on the LCA-group G induces, by convolution, a strongly continuous contraction semigroup $(P_t)_{t>0}$ on several Banach spaces of functions on G, e.g. $L^2(G)$ and $C_0(G)$. After a characterization of the strongly continuous contraction semigroups which arise in this way, such semigroups will be studied, and it turns out that important properties of the semigroups are equivalent with certain properties of the continuous negative definite function ψ on the dual group Γ associated with $(\mu_t)_{t>0}$. Finally, the infinitesimal generator, the potential operator and the zero-resolvent for the induced semigroup on $L^2(G)$ will be represented in terms of ψ.

12.1. Definition. A positive linear mapping N of $C_c(G)$ into $C(G)$ will be called a *continuous kernel* on G. A continuous kernel N on G is said to be *translation invariant*, if it commutes with the translations of G, i.e. if

$$N(\tau_a f)=\tau_a(Nf) \quad \text{for all } a\in G \text{ and } f\in C_c(G).$$

Let μ be a positive measure on G. The mapping $N\colon C_c(G)\to C(G)$ defined by

$$Nf(x)=\mu * f(x) \quad \text{for } f\in C_c(G) \text{ and } x\in G, \tag{1}$$

is clearly a translation invariant, continuous kernel on G, the *convolution kernel defined by* μ. Conversely, any translation invariant continuous kernel on G is given in this way:

12.2. Lemma. *Let N be a translation invariant continuous kernel on G. There exists a uniquely determined positive measure μ on G such that N is given in terms of μ by* (1).

Proof. Suppose μ is a positive measure on G such that (1) holds. Then

$$\langle\mu, f\rangle = N(\check{f})(0) \quad \text{for all } f\in C_c(G), \tag{2}$$

and in particular μ is uniquely determined. The mapping

$$f\mapsto N(\check{f})(0)$$

is clearly a positive linear form on $C_c(G)$, i.e. a positive measure μ on G, and since μ satisfies (2) we find that

$$Nf(x)=\tau_{-x}(Nf)(0)=N(\tau_{-x}f)(0)=\langle\mu, (\tau_{-x}f)^{\vee}\rangle=\mu * f(x)$$

for $f\in C_c(G)$ and $x\in G$. □

12.3. A positive bounded measure μ on G defines for $p\in[1,\infty[$ a bounded operator T_μ of norm $\leq \mu(G)$ on $L^p(G)$, by

$$T_\mu f=\mu * f \quad \text{for } f\in L^p(G).$$

The operator T_μ is called the *convolution operator* induced by μ, and it clearly satisfies

$$T_\mu(\tau_a f)=\tau_a(T_\mu f) \quad \text{for } a\in G \text{ and } f\in L^p(G) \tag{3}$$

and

$$0\leqq T_\mu f\leqq \mu(G) \quad \text{a.e. for all } f\in L^p(G) \text{ such that } 0\leqq f\leqq 1 \text{ a.e.} \tag{4}$$

It is an immediate consequence of the following lemma, that the properties (3) and (4) are characteristic for convolution operators induced by positive bounded measures.

12.4. Lemma. *Let T be a bounded operator on $L^p(G)$, $p\in[1,\infty[$, with the properties*

(i) *T is* translation invariant *i.e.*

$$T(\tau_a f)=\tau_a(Tf) \quad \textit{for } a\in G \textit{ and } f\in L^p(G),$$

(ii) *T is* submarkovian *i.e.*

$$f\in L^p(G) \quad \textit{and} \quad 0\leqq f\leqq 1 \quad \textit{a.e. implies } 0\leqq Tf\leqq 1 \textit{ a.e.}$$

Then there exists a uniquely determined positive bounded measure μ on G with total mass $\mu(G)\leqq 1$, such that

$$Tf=\mu * f \quad \textit{for all } f\in L^p(G). \tag{5}$$

Proof. Suppose that μ is a positive bounded measure on G such that (5) holds. Then Tf has a uniquely determined continuous representative for all $f\in C_c(G)$ and furthermore

$$\langle\mu, f\rangle=\mu*\check{f}(0)=T(\check{f})(0) \quad \text{for } f\in C_c(G),$$

where $T(\check{f})$ stands for the continuous representative for $T\check{f}$. This shows that μ is uniquely determined.

For the construction of the measure μ, we first suppose that Tf has a continuous representative for all $f\in C_c(G)$. It is then easy to see that the "restriction" of T to $C_c(G)$ is a translation invariant continuous kernel on G, and there exists, by Lemma 12.2, a positive measure μ on G such that

$$Tf=\mu*f \quad \text{for } f\in C_c(G).$$

By (ii) the measure μ is bounded with total mass $\mu(G)\leqq 1$, and since $C_c(G)$ is dense in $L^p(G)$, the equality $Tf=\mu*f$ extends to $L^p(G)$, thus yielding (5).

In the general case we consider the family $(T^V)_{V\in\check{\mathscr{V}}}$ of bounded operators on $L^p(G)$ defined by

$$T^V f=(Tf)*\varphi_V \quad \text{for } f\in L^p(G) \text{ and } V\in\check{\mathscr{V}},$$

where $(\varphi_V)_{V\in\check{V}}$ is some approximate unit. It is clear that each of the operators T^V is translation invariant and submarkovian and $T^V f$ has a continuous representative for every $f\in C_c(G)$. It follows from the discussion of the first case, that there exists for every $V\in\check{V}$ a positive bounded measure μ^V on G with total mass $\mu^V(G)\leq 1$ such that

$$T^V f=\mu^V * f \quad \text{for } f\in L^p(G).$$

Let μ denote a vague accumulation point for the net $(\mu^V)_{V\in\check{V}}$ as V "tends" to $\{0\}$. By Proposition 1.18 and Exercise 1.20 we then get

$$\mu * f=Tf \qquad \text{for } f\in C_c(G)$$

and, again by the density of $C_c(G)$ in $L^p(G)$, this implies (5). □

12.5. Remark. A convolution operator T_μ on $L^2(G)$ induced by a positive bounded measure μ on G is unitarily equivalent with the *multiplication operator* $M_{\hat{\mu}}$ on $L^2(\Gamma)$ defined by

$$M_{\hat{\mu}} g=\hat{\mu} g \quad \text{for } g\in L^2(\Gamma).$$

In fact, the Fourier transformation $\mathscr{F}_G$ is an isometry of $L^2(G)$ onto $L^2(\Gamma)$ (cf. Theorem 2.5) and furthermore

$$\mathscr{F}_G(T_\mu f)=M_{\hat{\mu}}(\mathscr{F}_G f) \quad \text{for } f\in L^2(G),$$

i.e. the following diagram is commutative

$$\begin{array}{ccc} L^2(G) & \xrightarrow{T_\mu} & L^2(G) \\ \big\downarrow{\scriptstyle \mathscr{F}_G} & & \big\downarrow{\scriptstyle \mathscr{F}_G} \\ L^2(\Gamma) & \xrightarrow{M_{\hat{\mu}}} & L^2(\Gamma). \end{array}$$

12.6. Exercise. Let μ be a positive bounded measure on G with induced convolution operator T_μ on $L^2(G)$. The spectrum $\sigma(T_\mu)$ of T_μ is given by

$$\sigma(T_\mu)=\overline{\{\hat{\mu}(\gamma)\mid \gamma\in\Gamma\}}.$$

We have $(T_\mu)^*=T_{\check{\mu}}$ and T_μ is unitary if and only if μ is equal to ε_a for some $a\in G$. The operator T_μ is positive if and only if μ is positive definite.

12.7. Proposition. *Let $p\in[1,\infty[$. There is a one-to-one correspondence between convolution semigroups $(\mu_t)_{t>0}$ on G and strongly continuous contraction semigroups of translation invariant and submarkovian operators $(P_t)_{t>0}$ on $L^p(G)$; $(\mu_t)_{t>0}$ and $(P_t)_{t>0}$ are corresponding if and only if*

$$P_t f=\mu_t * f \quad \textit{for } f\in L^p(G) \textit{ and } t>0. \tag{6}$$

Proof. Let $(\mu_t)_{t>0}$ be a convolution semigroup on G. It is clear that the family $(P_t)_{t>0}$ of operators on $L^p(G)$ defined by (6) is a contraction semigroup and that all the operators P_t are translation invariant and submarkovian (cf. 12.3). To see that $(P_t)_{t>0}$ is strongly continuous we may (and do) assume that all the measures μ_t are probability measures (cf. Corollary 8.6). Let $f \in L^p(G)$. Using Hölder's inequality for the measure μ_t we find

$$\begin{aligned}\|P_t f - f\|_p^p &= \int \left| \int (\tau_y f(x) - f(x))\, d\mu_t(y)\right|^p dx \\ &\leqq \int \left(\int |\tau_y f(x) - f(x)|^p \, d\mu_t(y)\right) dx \\ &= \int \|\tau_y f - f\|_p^p \, d\mu_t(y).\end{aligned}$$

Since the function $y \mapsto \|\tau_y f - f\|_p^p$ is bounded and continuous (cf. Rudin [1], p. 3), Proposition 8.2 gives that

$$\lim_{t\to 0} \|P_t f - f\|_p^p = 0.$$

Let now conversely $(P_t)_{t>0}$ be a strongly continuous contraction semigroup of translation invariant and submarkovian operators on $L^p(G)$. By Lemma 12.4 there exists for every $t>0$ a uniquely determined positive bounded measure μ_t on G with total mass $\mu_t(G) \leqq 1$, such that

$$P_t f = \mu_t * f \quad \text{for } f \in L^p(G).$$

The family $(\mu_t)_{t>0}$ is the desired convolution semigroup on G. For $t, s > 0$ and $f \in C_c(G)$ we find

$$\langle \mu_{t+s}, f\rangle = P_{t+s}(\check{f})(0) = P_t(P_s(\check{f}))(0) = \mu_t * (\mu_s * \check{f})(0) = \langle \mu_t * \mu_s, f\rangle.$$

(We may choose continuous representatives for the L^p-functions in question.) Finally, for $f, g \in C_c(G)$ and $t>0$ we have

$$\langle \mu_t, f * g\rangle = \int \mu_t * \check{f}(x)\, g(x)\, dx = \int P_t(\check{f})(x)\, g(x)\, dx,$$

and by the strong continuity of $(P_t)_{t>0}$ this gives

$$\lim_{t\to 0} \langle \mu_t, f * g\rangle = \int \check{f}(x)\, g(x)\, dx = f * g(0).$$

It is easy to see that this implies

$$\lim_{t\to 0} \mu_t = \varepsilon_0 \quad \text{vaguely.} \quad \square$$

When $(\mu_t)_{t>0}$ and $(P_t)_{t>0}$ correspond to each other by the above correspondence, $(P_t)_{t>0}$ will be called the semigroup *induced* by $(\mu_t)_{t>0}$.

12.8. Definition. A strongly continuous contraction semigroup $(P_t)_{t>0}$ on $C_0(G)$, for which all the operators P_t are positive, i.e. such that for all $t>0$

$$f \in C_0^+(G) \quad \text{implies} \quad P_t f \in C_0^+(G),$$

is called a *Feller semigroup* on G. A Feller semigroup $(P_t)_{t>0}$ on G is said to be *translation invariant* if all the operators P_t commute with the translations of G, i.e. if

$$P_t(\tau_a f) = \tau_a(P_t f) \quad \text{for } a \in G,\ t>0 \text{ and } f \in C_0(G).$$

12.9. A convolution semigroup $(\mu_t)_{t>0}$ on G *induces* a family $(P_t)_{t>0}$ of operators on $C_0(G)$ by the definition

$$P_t f = \mu_t * f \quad \text{for } f \in C_0(G) \text{ and } t>0, \tag{7}$$

and it is easy to see that $(P_t)_{t>0}$ is a translation invariant Feller semigroup on G (for the strong continuity, see Exercise 1.21). Conversely, every translation invariant Feller semigroup on G can be obtained in this way:

12.10. Exercise. Let $(P_t)_{t>0}$ be a translation invariant Feller semigroup on G. Then there exists a uniquely determined convolution semigroup $(\mu_t)_{t>0}$ on G such that $(P_t)_{t>0}$ is given in terms of $(\mu_t)_{t>0}$ by (7).

12.11. Exercise. Let $(\mu_t)_{t>0}$ be a convolution semigroup on G and let $(P_t)_{t>0}$ denote the induced semigroup on E, where E is any one of the spaces $L^p(G)$ for $p \in [1, \infty[$ or $C_0(G)$. The resolvent $(N_\lambda)_{\lambda>0}$ for $(P_t)_{t>0}$ on E (cf. 11.11) is given by

$$N_\lambda f = \rho_\lambda * f \quad \text{for } f \in E \text{ and } \lambda>0,$$

where $(\rho_\lambda)_{\lambda>0}$ is the resolvent for the convolution semigroup $(\mu_t)_{t>0}$ (cf. 8.12).

Let $(B, D(B))$ denote either the infinitesimal generator $(A, D(A))$ for $(P_t)_{t>0}$ on E or the potential operator $(N, D(N))$ for $(P_t)_{t>0}$ on E, or the zero-resolvent $(N_0, D(N_0))$ for $(P_t)_{t>0}$ on E.

The operator $(B, D(B))$ is translation invariant in the sense that

(i) $\tau_x(D(B)) = D(B)$ for all $x \in G$,

(ii) $\tau_x(Bf) = B(\tau_x f)$ for all $x \in G$ and $f \in D(B)$,

and has the following property:

(iii) For $f \in D(B)$ and $g \in C_c(G)$ the function $f * g$ belongs to $D(B)$ and satisfies

$$B(f*g) = (Bf)*g.$$

Let $(\mu_t)_{t>0}$ be a convolution semigroup on G with associated continuous negative definite function ψ on Γ. We shall obtain necessary and sufficient conditions on ψ under which the zero-resolvent, respectively the potential operator, for the induced semigroup on $L^2(G)$ or $C_0(G)$ is densely defined.

12.12. Theorem. *Let $(P_t)_{t>0}$ be the strongly continuous contraction semigroup on $L^2(G)$ induced by $(\mu_t)_{t>0}$.*

(i) The zero resolvent $(N_0, D(N_0))$ for $(P_t)_{t>0}$ is densely defined if and only if $\psi \neq 0$ locally a.e. on Γ.

(ii) The potential operator $(N, D(N))$ for $(P_t)_{t>0}$ is densely defined if and only if $\operatorname{Re}\psi \neq 0$ locally a.e. on Γ.

Proof. Let $(N_\lambda)_{\lambda>0}$ be the resolvent for $(P_t)_{t>0}$ on $L^2(G)$ and $(\rho_\lambda)_{\lambda>0}$ the resolvent for $(\mu_t)_{t>0}$.

We shall first consider the case $\psi(0)>0$. The norm of the operator P_s on $L^2(G)$ satisfies

$$\|P_s\|_2 \leqq \mu_s(G) = \exp(-s\psi(0)) \quad \text{for } s>0,$$

and this gives for $f \in L^2(G)$ that

$$\int_0^\infty \|P_s f\|_2 \, ds \leqq \frac{\|f\|_2}{\psi(0)}.$$

It follows that the integral $\int_0^\infty P_s f\, ds$ exists in $L^2(G)$ for all $f \in L^2(G)$, and this gives for $f \in L^2(G)$ that

$$\lim_{\lambda\to 0} N_\lambda f = \int_0^\infty P_s f\, ds$$

and

$$\lim_{t\to\infty} \int_0^t P_s f\, ds = \int_0^\infty P_s f\, ds.$$

The zero-resolvent and the potential operator for $(P_t)_{t>0}$ are thus everywhere defined and bounded operators on $L^2(G)$.

For the rest of the proof we suppose that $\psi(0)=0$.

(i) The set

$$\Gamma_0 = \{\gamma \in \Gamma \mid \psi(\gamma) = 0\}$$

is a closed subgroup of Γ, cf. 8.27.

For $\varphi \in C_c(\Gamma)$ we have

$$\mathscr{F}_G(\lambda N_\lambda(\bar{\mathscr{F}}_\Gamma \varphi)) = \frac{\lambda}{\lambda+\psi}\varphi \quad \text{for } \lambda>0,$$

and if Γ_0 is locally negligible, the dominated convergence theorem gives that

$$\lim_{\lambda\to 0} \|\lambda N_\lambda(\bar{\mathscr{F}}_\Gamma \varphi)\|_2^2 = \lim_{\lambda\to 0} \int \left| \frac{\lambda}{\lambda+\psi(\gamma)} \varphi(\gamma) \right|^2 d\gamma = 0.$$

Since the set $\{\bar{\mathscr{F}}_\Gamma \varphi \mid \varphi \in C_c(\Gamma)\}$ is dense in $L^2(G)$, and the set $\{f \in L^2(G) \mid \lim_{\lambda\to 0} \lambda N_\lambda f = 0\}$ is closed, we conclude that condition (iii) of Proposition 11.14 is fulfilled, and N_0 is thus densely defined.

If Γ_0 is not locally negligible, then Γ_0 is open (cf. Bourbaki [3] p. 169). The measures μ_t are concentrated on the compact subgroup $\Gamma_0^\perp$ of G (cf. Proposition 8.27 and 2.9), and it follows that the measures $(\lambda\rho_\lambda)*(\lambda\check{\rho}_\lambda)$ are concentrated on $\Gamma_0^\perp$. Choosing $f\in C_c^+(G)$ such that $f*\check{f}\geqq 1$ on $\Gamma_0^\perp$ we find for all $\lambda>0$,

$$1\leqq\langle\lambda\rho_\lambda*\lambda\check{\rho}_\lambda, f*\check{f}\rangle=\int|\lambda\rho_\lambda*f(x)|^2dx=\|\lambda N_\lambda f\|_2^2,$$

and condition (iii) of Proposition 11.14 is thus not fulfilled, i.e. $D(N_0)$ is not dense in $L^2(G)$.

(ii) The set

$$\Gamma_1=\{\gamma\in\Gamma\,|\,\operatorname{Re}\psi(\gamma)=0\}$$

is a closed subgroup of Γ.

For $\varphi\in C_c(\Gamma)$ we have

$$\mathscr{F}_G(P_t(\bar{\mathscr{F}}_\Gamma\varphi))=\exp(-t\psi)\varphi \qquad \text{for } t>0,$$

and if Γ_1 is locally negligible, we find as above

$$\lim_{t\to\infty}\|P_t(\bar{\mathscr{F}}_\Gamma\varphi)\|_2^2=\lim_{t\to\infty}\int\exp(-2t\operatorname{Re}\psi(\gamma))|\varphi(\gamma)|^2d\gamma=0,$$

and this implies by Proposition 11.9 that N is densely defined.

If Γ_1 is not locally negligible, then Γ_1 is open and the measures $\mu_t*\check{\mu}_t$ are concentrated on the compact subgroup $\Gamma_1^\perp$ of G. Choosing $f\in C_c^+(G)$ such that $f*\check{f}\geqq 1$ on $\Gamma_1^\perp$ we find for all $t>0$

$$1\leqq\langle\mu_t*\check{\mu}_t, f*\check{f}\rangle=\int|\mu_t*f(x)|^2dx=\|P_tf\|_2^2,$$

and it follows that $D(N)$ is not dense in $L^2(G)$. □

12.13. Exercise. Let $(P_t)_{t>0}$ be the Feller semigroup on G induced by $(\mu_t)_{t>0}$.

1) The zero-resolvent for $(P_t)_{t>0}$ is densely defined if and only if $\psi\neq 0$ locally a.e. on Γ.

2) The potential operator for $(P_t)_{t>0}$ is densely defined if and only if $\operatorname{Re}\psi\neq 0$ locally a.e. on Γ.

12.14. Remark. 1) If $\psi(0)>0$ then the zero-resolvent $(N_0, D(N_0))$ on $L^2(G)$ (resp. on $C_0(G)$) is equal to the potential operator $(N, D(N))$ on $L^2(G)$ (resp. $C_0(G)$), and $N_0=N$ is an everywhere defined and bounded operator. (Cf. Exercise 12.22.)

2) If G is compact, then the statements: "$\psi\neq 0$ locally a.e. on Γ", "$\operatorname{Re}\psi\neq 0$ locally a.e. on Γ" and "$\psi(0)>0$" are equivalent.

It is clear by Proposition 11.14 that the zero-resolvent for the semigroup induced by the degenerate convolution semigroup $(\varepsilon_0)_{t>0}$ is not densely defined. Likewise, by Proposition 11.9, it is clear that the potential operator for the semigroup induced by a translation semigroup on G is not densely defined.

If G has no compact subgroups except $\{0\}$, then these cases are the only cases, where the zero-resolvent and the potential operator are not densely defined. (G has no compact subgroups except $\{0\}$ if and only if Γ is connected, cf. Bourbaki [4] p. 136, but we do not need this result.)

12.15. Corollary. *Suppose that G has no compact subgroups except $\{0\}$, and let $(P_t)_{t>0}$ be the semigroup on $L^2(G)$ or $C_0(G)$ induced by $(\mu_t)_{t>0}$. Then we have:*

(i) *The zero-resolvent for $(P_t)_{t>0}$ is densely defined except when $\mu_t = \varepsilon_0$ for all $t>0$.*

(ii) *The potential operator for $(P_t)_{t>0}$ is densely defined except when $(\mu_t)_{t>0}$ is a translation semigroup.*

Proof. (i) If the zero-resolvent for $(P_t)_{t>0}$ is not densely defined, then by 12.12 and 12.13 the set $\Gamma_0 = \{\gamma \in \Gamma \mid \psi(\gamma) = 0\}$ is an open subgroup of Γ. The compact subgroup $\Gamma_0^{\perp}$ of G must be $\{0\}$ by hypothesis and hence $\Gamma_0 = \Gamma$. It follows that ψ is identically zero and hence $\mu_t = \varepsilon_0$ for $t>0$.

(ii). If the potential operator for $(P_t)_{t>0}$ is not densely defined, then the set $\Gamma_1 = \{\gamma \in \Gamma \mid \operatorname{Re}\psi(\gamma) = 0\}$ is an open subgroup of Γ, and hence $\Gamma_1 = \Gamma$. The function ψ is thus purely imaginary, and by Proposition 7.20 and 8.11 we get that $(\mu_t)_{t>0}$ is a translation semigroup. □

Let $(\mu_t)_{t>0}$ be a convolution semigroup on G with associated continuous negative definite function ψ on Γ, and let $(P_t)_{t>0}$ be the induced semigroup on $L^2(G)$.

In the next sections we will express the infinitesimal generator, the zero-resolvent and the potential operator in terms of ψ.

12.16. Theorem. *The infinitesimal generator $(A, D(A))$ for $(P_t)_{t>0}$ is given by*

$$D(A) = \{f \in L^2(G) \mid \psi \mathscr{F}_G f \in L^2(\Gamma)\},$$

and

$$\mathscr{F}_G(Af) = -\psi \mathscr{F}_G f \quad \text{for } f \in D(A).$$

For every $\varphi \in C_c(\Gamma)$ we have $\overline{\mathscr{F}}_\Gamma \varphi \in D(A)$.

Proof. For $f \in D(A)$ we have

$$\lim_{t\to 0} \frac{1}{t}(P_t f - f) = Af \quad \text{in } L^2(G),$$

or by the Plancherel theorem

$$\lim_{t\to 0} \mathscr{F}_G\left(\frac{1}{t}(P_t f - f)\right) = \mathscr{F}_G(Af) \quad \text{in } L^2(\Gamma).$$

It follows that there exists a sequence $(t_n)_{n\in\mathbb{N}}$ such that $t_n \to 0$ and such that

$$\lim_{n\to\infty} \left[\frac{1}{t_n}(e^{-t_n\psi} - 1)\mathscr{F}_G f\right] = \mathscr{F}_G(Af) \quad \text{a.e. on } \Gamma.$$

However, for $\gamma \in \Gamma$ we have

$$\lim_{n\to\infty} \frac{1}{t_n}(e^{-t_n\psi(\gamma)}-1) = -\psi(\gamma),$$

and it follows that

$$\mathscr{F}_G(Af) = -\psi\,\mathscr{F}_G f \quad \text{a.e. on } \Gamma,$$

in particular $\psi\,\mathscr{F}_G f \in L^2(\Gamma)$ and $\mathscr{F}_G(Af) = -\psi\,\mathscr{F}_G f$ in $L^2(\Gamma)$.

Suppose next that $f \in L^2(G)$ and $\psi\,\mathscr{F}_G f \in L^2(\Gamma)$. Since

$$\lim_{t\to 0} \frac{1}{t}(e^{-t\psi}-1)\mathscr{F}_G f = -\psi\,\mathscr{F}_G f \quad \text{a.e. on } \Gamma,$$

and furthermore

$$\left|\frac{1}{t}(e^{-t\psi(\gamma)}-1)\right| \leqq |\psi(\gamma)| \quad \text{for } \gamma \in \Gamma,$$

by the elementary inequality

$$|e^{-z}-1| \leqq |z|,$$

valid for $z \in \mathbb{C}$ with $\operatorname{Re} z \geqq 0$, the dominated convergence theorem gives that

$$\lim_{t\to 0} \mathscr{F}_G\left(\frac{1}{t}(P_t f - f)\right) = \lim_{t\to 0} \frac{1}{t}(e^{-t\psi}-1)\mathscr{F}_G f = -\psi\,\mathscr{F}_G f$$

in $L^2(\Gamma)$, hence by the Plancherel theorem that

$$\lim_{t\to 0} \frac{1}{t}(P_t f - f) \quad \text{exists in } L^2(G),$$

i.e. $f \in D(A)$. The last assertion is clear by the characterization of $D(A)$. □

12.17. Remark. The infinitesimal generator $(A, D(A))$ is unitarily equivalent, via the Fourier transformation (cf. 12.5), with the multiplication operator $M_{-\psi}$ on $L^2(\Gamma)$ whose domain is

$$D(M_{-\psi}) = \{g \in L^2(\Gamma) \mid \psi g \in L^2(\Gamma)\}$$

and defined by

$$M_{-\psi} g = -\psi g \quad \text{for } g \in D(M_{-\psi}).$$

12.18. Corollary. *The infinitesimal generator* $(A, D(A))$ *is injective if and only if the zero-resolvent* $(N_0, D(N_0))$ *is densely defined.*

Proof. Suppose that $(N_0, D(N_0))$ is not densely defined. By Theorem 12.12 it follows that the set $\{\gamma\in\Gamma \mid \psi(\gamma)=0\}$ is an open subgroup of Γ, and we can therefore choose a function $\varphi\in C_c^+(\Gamma)$ such that $\varphi\neq 0$ and $\psi\varphi=0$. The function $\bar{\mathscr{F}}_\Gamma\varphi$ belongs to $D(A)$ and $\bar{\mathscr{F}}_\Gamma\varphi\neq 0$, while $\mathscr{F}_G(A(\bar{\mathscr{F}}_\Gamma\varphi))=-\psi\varphi=0$, and A is consequently not injective.

The "if"-part follows from Proposition 11.14. □

12.19. Corollary. *Suppose that the zero-resolvent* $(N_0, D(N_0))$ *is densely defined. Then the function* $1/\psi$ *is defined locally a.e. on* Γ *and the zero-resolvent is determined by*

$$D(N_0)=\left\{f\in L^2(G)\,\middle|\,\frac{1}{\psi}\mathscr{F}_G f\in L^2(\Gamma)\right\}$$

and

$$\mathscr{F}_G(N_0 f)=\frac{1}{\psi}\mathscr{F}_G f \quad \textit{for } f\in D(N_0).$$

Proof. By Theorem 12.12 the function $1/\psi$ is defined locally a.e. on Γ. We first remark that the expression $\frac{1}{\psi}\mathscr{F}_G f\in L^2(\Gamma)$ means that there exists an element $g\in L^2(\Gamma)$ such that the function $\frac{1}{\psi}\mathscr{F}_G f$ (which is defined locally a.e. on Γ) coincides locally a.e. on Γ with g. Such a function $g\in L^2(\Gamma)$ is uniquely determined. For $f\in D(N_0)$ we get by Proposition 11.14 that

$$-f=A(N_0 f)$$

or by Theorem 12.16

$$-\mathscr{F}_G f=\mathscr{F}_G(A(N_0 f))=-\psi\mathscr{F}_G(N_0 f),$$

and it follows that

$$\mathscr{F}_G(N_0 f)=\frac{1}{\psi}\mathscr{F}_G f\in L^2(\Gamma).$$

Let $f\in L^2(G)$ and suppose that there exists $g\in L^2(\Gamma)$ such that $g=\frac{1}{\psi}\mathscr{F}_G f$ locally a.e. on Γ. For $h\in L^2(G)$ with $\mathscr{F}_G h=g$ we have $\mathscr{F}_G f=\psi\mathscr{F}_G h$ locally a.e. on Γ. The function $\psi\mathscr{F}_G h$ is defined a.e. on Γ, and since ψ is continuous and $|\mathscr{F}_G h|^2$ is integrable we have

$$\begin{aligned}\int|\psi\mathscr{F}_G h|^2\,d\gamma&=\sup_\varphi\int|\psi\mathscr{F}_G h|^2\varphi\,d\gamma\\&=\sup_\varphi\int|\mathscr{F}_G f|^2\varphi\,d\gamma=\|\mathscr{F}_G f\|_2^2<\infty,\end{aligned}$$

where the suprema are taken over the set $\{\varphi\in C_c(\Gamma)\mid 0\leqq\varphi\leqq 1\}$. This shows that $\psi\mathscr{F}_G h\in L^2(\Gamma)$ and hence $\mathscr{F}_G f=\psi\mathscr{F}_G h$ a.e. on Γ, which by 12.16 implies that

$h \in D(A)$ and

$$\mathscr{F}_G(Ah) = -\psi \mathscr{F}_G h = -\mathscr{F}_G f,$$

and consequently

$$f = -Ah \in R(A) = D(N_0). \quad \square$$

12.20. Remark. When the zero-resolvent $(N_0, D(N_0))$ is densely defined, it is unitarily equivalent, via the Fourier transformation (cf. 12.5), with the multiplication operator $M_{1/\psi}$ on $L^2(\Gamma)$.

12.21. Remark. The analogous result for the potential operator $(N, D(N))$ is clearly contained in Corollary 12.19, because if N is densely defined then N_0 and N are equal (cf. Proposition 11.15).

12.22. Exercise. If the zero-resolvent $(N_0, D(N_0))$ for the induced semigroup $(P_t)_{t>0}$ on $L^2(G)$ or $C_0(G)$ is everywhere defined, then $\psi(0) > 0$.

12.23. Exercise. The family $(P_t^*)_{t>0}$ of adjoint operators is a contraction semigroup on $L^2(G)$ induced by the convolution semigroup $(\check{\mu}_t)_{t>0}$ on G, and the infinitesimal generator for $(P_t^*)_{t>0}$ is the operator $(A^*, D(A^*))$, adjoint to $(A, D(A))$. The following conditions are equivalent:

(i) $(A, D(A))$ is self-adjoint.

(ii) P_t is self-adjoint for $t>0$.

(iii) ψ is a real function on Γ.

12.24. Exercise. 1) The infinitesimal generator $(A, D(A))$ on $L^2(G)$ is a normal operator whose spectrum is given by

$$\sigma(A) = \overline{\{-\psi(\gamma) \mid \gamma \in \Gamma\}}.$$

2) The infinitesimal generator $(A, D(A))$ is injective if and only if the range $R(A)$ of A is dense in $L^2(G)$.

12.25. Example. Let $(\mu_t)_{t>0}$ be the Brownian semigroup on $\mathbb{R}^n (n \geq 1)$, cf. 10.3, and let $(P_t)_{t>0}$ denote the induced semigroup on $C_0(\mathbb{R}^n)$.

The continuous negative definite function on $\mathbb{R}^n$ associated with $(\mu_t)_{t>0}$ is $y \mapsto \|y\|^2$, and it follows by Exercise 12.13 that the potential operator $(N, D(N))$ for $(P_t)_{t>0}$ is densely defined (and equal to the zero-resolvent for $(P_t)_{t>0}$) for all $n \geq 1$.

It seems to be difficult to obtain a precise description of $D(N)$, but it is possible to exhibit "rich" subsets of $D(N)$, cf. Sato [1]:

$n=1$. A function $f \in C_c(\mathbb{R})$ belongs to $D(N)$ if and only if

$$\int_{\mathbb{R}} f(y)\,dy = \int_{\mathbb{R}} y f(y)\,dy = 0,$$

and

$$Nf(x) = -\frac{1}{2} \int_{\mathbb{R}} f(y)|x-y|\,dy \qquad \text{for } f \in C_c(\mathbb{R}) \cap D(N).$$

$n=2$. A function $f\in C_c(\mathbb{R}^2)$ belongs to $D(N)$ if and only if

$$\int_{\mathbb{R}^2} f(y)\,dy=0,$$

and

$$Nf(x)=-\frac{1}{2\pi}\int_{\mathbb{R}^2} f(y)\log\|x-y\|\,dy \qquad \text{for } f\in C_c(\mathbb{R}^2)\cap D(N).$$

$n\geqq 3$. The set $C_c(\mathbb{R}^n)$ is contained in $D(N)$ and

$$Nf(x)=\frac{\Gamma\left(\frac{n}{2}\right)}{2(n-2)\pi^{n/2}}\int_{\mathbb{R}^n} f(y)\|x-y\|^{2-n}\,dy \qquad \text{for } f\in C_c(\mathbb{R}^n),$$

cf. 13.29.

For a representation of the infinitesimal generator see Example 18.31.

12.26. Remark. Proposition 12.7 and the analogous result in Exercise 12.10 give the connection to some more specialized (translation invariant) potential theories.

The set of translation invariant Dirichlet forms on G is in one-to-one correspondence with the set of strongly continuous contraction semigroups on $L^2(G)$ which consist of translation invariant, sub-markovian and self-adjoint operators, hence (cf. Exercise 12.23) in one-to-one correspondence with the set of real and continuous negative definite functions on Γ, cf. Deny [6], and Berg and Forst [1] for an extension to certain non-symmetric translation invariant Dirichlet forms.

A translation invariant probabilistic potential theory on G is determined by its semigroup of transition probabilities (cf. 8.29), or equivalently by a translation invariant Feller semigroup on G.

Properties of translation invariant Feller semigroups were studied in terms of the "associated" continuous negative definite function in Berg [1], where the C_0-version of 12.12 was proved.

For further examples of potential operators we refer to Sato [1].

Chapter III. Potential Theory for Transient Convolution Semigroups

§ 13. Transient Convolution Semigroups

Let $(\mu_t)_{t>0}$ be a convolution semigroup on the LCA-group G, and let $(P_t)_{t>0}$ denote the induced Feller semigroup. It is often useful, in the study of $(\mu_t)_{t>0}$, to replace the potential operator $N=\int_0^\infty P_s ds$ for $(P_t)_{t>0}$ with the measure (if it exists)

$$\kappa=\int_0^\infty \mu_t\, dt.$$

If this measure exists, then the convolution semigroup $(\mu_t)_{t>0}$ is said to be transient and κ is called the potential kernel for $(\mu_t)_{t>0}$. If $C_c(G)$ is contained in the domain of N, then $(\mu_t)_{t>0}$ is said to be integrable, and it is shown that this is the case if and only if $(\mu_t)_{t>0}$ is transient and the potential kernel κ tends to zero at infinity.

The properties of transience and integrability of a convolution semigroup $(\mu_t)_{t>0}$ on G are discussed in terms of the associated continuous negative definite function ψ on the dual group Γ, and the theory is illustrated with many concrete examples.

It turns out that the potential kernel κ for a transient convolution semigroup is "almost" positive definite, in the sense that $\frac{1}{2}(\kappa+\check{\kappa})$ is positive definite, and the paragraph is finished with an "explicit" calculation of the Fourier transform of $\frac{1}{2}(\kappa+\check{\kappa})$ in terms of ψ.

13.1. Let $(\mu_t)_{t>0}$ be a convolution semigroup on G with resolvent $(\rho_\lambda)_{\lambda>0}$, cf. 8.12. For $f\in C_c^+(G)$ the mapping

$$\lambda\mapsto\langle\rho_\lambda, f\rangle=\int_0^\infty e^{-\lambda t}\langle\mu_t, f\rangle dt$$

of $]0,\infty[$ into $\mathbb{R}$ is decreasing, and by the monotone convergence theorem we find

$$\lim_{\lambda\to 0}\langle\rho_\lambda, f\rangle=\int_0^\infty\langle\mu_t, f\rangle dt\leqq\infty.$$

If $\lim_{\lambda\to 0}\langle\rho_\lambda, f\rangle$ is finite for all $f\in C_c^+(G)$, a positive measure κ on G is defined by

$$\langle\kappa, f\rangle=\lim_{\lambda\to 0}\langle\rho_\lambda, f\rangle=\int_0^\infty\langle\mu_t, f\rangle dt \qquad \text{for } f\in C_c^+(G), \tag{1}$$

and the measure κ is in particular vague limit for the measures ρ_λ as λ tends to zero.

13.2. Definition. The convolution semigroup $(\mu_t)_{t>0}$ on G is said to be *transient* if

$$\lim_{\lambda\to 0}\langle\rho_\lambda, f\rangle=\int_0^\infty\langle\mu_t, f\rangle\,dt<\infty \qquad \text{for all } f\in C_c^+(G),$$

and the measure κ defined by (1) is called the *potential kernel* for $(\mu_t)_{t>0}$.

The convolution semigroup $(\mu_t)_{t>0}$ is said to be *recurrent* if $(\mu_t)_{t>0}$ is not transient.

A measure κ on G will be called a *potential kernel*, if κ is potential kernel for some transient convolution semigroup on G. It follows from Proposition 15.21 that a potential kernel is potential kernel for exactly one transient convolution semigroup.

Let $(\mu_t)_{t>0}$ be a transient convolution semigroup on G with potential kernel κ. It is easy to see that

$$\langle\kappa, f\rangle=\lim_{\lambda\to 0}\langle\rho_\lambda, f\rangle=\sup_{\lambda>0}\langle\rho_\lambda, f\rangle=\int_0^\infty\langle\mu_t, f\rangle\,dt\leqq\infty, \tag{2}$$

for all lower semicontinuous functions $f\colon G\to[0,\infty]$.

13.3. Proposition. *Let $(\mu_t)_{t>0}$ be a transient convolution semigroup on G with potential kernel κ and resolvent $(\rho_\lambda)_{\lambda>0}$.*

For $\lambda>0$ the convolution $\kappa*\rho_\lambda$ exists and we have

$$\kappa-\rho_\lambda=\lambda\kappa*\rho_\lambda,$$

i.e. the resolvent equation "holds" for $\mu=0$.

Furthermore we have

$$\operatorname{supp}(\kappa)=\operatorname{supp}(\rho_\lambda)=\overline{\bigcup_{t>0}\operatorname{supp}(\mu_t)} \qquad \text{for } \lambda>0,$$

and $\operatorname{supp}(\kappa)$ *is a semigroup in G.*

Proof. For $\lambda>0$ and $\mu\in\left]0,\frac{\lambda}{2}\right[$ we have

$$\rho_\mu*\rho_\lambda=\frac{1}{\lambda-\mu}(\rho_\mu-\rho_\lambda)\leqq\frac{2}{\lambda}\rho_\mu\leqq\frac{2}{\lambda}\kappa.$$

Let $f\in C_c^+(G)$. Since ρ_μ increases to κ as $\mu\searrow 0$ the family $\check\rho_\mu*f$ increases pointwise to $\check\kappa*f$ as $\mu\searrow 0$ and it follows by the monotone convergence theorem that

$$\int\int f(x+y)\,d\kappa(x)\,d\rho_\lambda(y)=\lim_{\mu\to 0}\langle\rho_\lambda,\check\rho_\mu*f\rangle\leqq\frac{2}{\lambda}\langle\kappa, f\rangle<\infty,$$

which shows that the convolution $\kappa*\rho_\lambda$ exists and moreover

$$\begin{aligned}\lambda\langle\kappa*\rho_\lambda, f\rangle&=\lim_{\mu\to 0}(\lambda-\mu)\langle\rho_\lambda,\check\rho_\mu*f\rangle\\&=\lim_{\mu\to 0}\langle\rho_\mu-\rho_\lambda, f\rangle=\langle\kappa-\rho_\lambda, f\rangle.\end{aligned}$$

By the formula

$$\langle\kappa, f\rangle = \int_0^\infty \langle\mu_t, f\rangle\, dt \qquad \text{for } f\in C_c(G)$$

it follows as in the proof of Theorem 8.24 that

$$\operatorname{supp}(\kappa) = \overline{\bigcup_{t>0} \operatorname{supp}(\mu_t)},$$

and the last part of the proposition follows by Theorem 8.24. □

13.4. Remark. Let ψ be the continuous negative definite function on Γ associated with the convolution semigroup $(\mu_t)_{t>0}$ on G.

If $\psi(0)>0$ it follows by Corollary 8.6 that $(\mu_t)_{t>0}$ is transient and the potential kernel κ for $(\mu_t)_{t>0}$ is a bounded measure with total mass $\kappa(G)=\dfrac{1}{\psi(0)}$. The Fourier transform $\hat{\kappa}$ of κ is easily seen to be

$$\hat{\kappa}(\gamma) = \frac{1}{\psi(\gamma)} \qquad \text{for } \gamma\in\Gamma. \tag{3}$$

On the other hand there exist transient as well as recurrent convolution semigroups $(\mu_t)_{t>0}$ consisting of probability measures (i.e. for which $\psi(0)=0$), cf. Example 13.29.

The potential kernel κ for a transient convolution semigroup of probability measures on G is an unbounded measure, and equation (3) has no meaning in this case. However, under additional assumptions, equation (3) can be generalized to the case $\psi(0)=0$, using the theory of Fourier transformation of positive definite measures, cf. 13.23 and 13.35.

A convolution semigroup $(\mu_t)_{t>0}$ on a compact group G, or more generally a convolution semigroup $(\mu_t)_{t>0}$ for which all the measures μ_t for $t>0$ are concentrated on a fixed compact set, is transient if and only if $\psi(0)>0$.

13.5. Lemma. *The convolution semigroup $(e^{-t}\exp(t\mu))_{t>0}$ determined by a positive bounded measure μ on G with total mass $\mu(G)\leqq 1$ is transient if and only if the series $\sum_{n=0}^{\infty}\mu^n$ is vaguely convergent, and the potential kernel is in the affirmative case given by*

$$\kappa = \sum_{n=0}^{\infty} \mu^n.$$

Proof. The resolvent $(\rho_\lambda)_{\lambda>0}$ for the convolution semigroup $(e^{-t}\exp(t\mu))_{t>0}$ is given in the following way:

$$\begin{aligned}
\rho_\lambda &= \int_0^\infty e^{-\lambda t} e^{-t} \exp(t\mu)\, dt \\
&= \int_0^\infty e^{-(\lambda+1)t}\left(\sum_{n=0}^{\infty} \frac{t^n}{n!}\mu^n\right) dt \\
&= \sum_{n=0}^{\infty} \mu^n \frac{1}{n!} \int_0^\infty t^n e^{-(\lambda+1)t}\, dt \\
&= \sum_{n=0}^{\infty} \left(\frac{1}{\lambda+1}\right)^{n+1} \mu^n,
\end{aligned}$$

and we have consequently for $f \in C_c^+(G)$ that

$$\lim_{\lambda \to 0} \langle \rho_\lambda, f \rangle = \sum_{n=0}^{\infty} \langle \mu^n, f \rangle \leqq \infty. \quad \square$$

13.6. Definition. A positive measure κ on G is called an *elementary kernel* if it has the form

$$\kappa = \sum_{n=0}^{\infty} \mu^n, \tag{4}$$

where μ is a positive bounded measure on G with total mass $\mu(G) \leqq 1$, such that the series $\sum_{n=0}^{\infty} \mu^n$ is vaguely convergent.

For an elementary kernel κ there exists by 15.22 exactly one positive bounded measure μ on G such that (4) holds. The elementary kernel (4) will be called the *elementary kernel determined by* μ.

The elementary kernel κ determined by μ is the potential kernel for the transient convolution semigroup $(\mu_t)_{t>0}$ on G determined by μ.

13.7. Proposition. *Let $(\mu_t)_{t>0}$ be a transient convolution semigroup on G with potential kernel κ, and let $(\rho_\lambda)_{\lambda>0}$ be the resolvent for $(\mu_t)_{t>0}$. The measure $\lambda\kappa + \varepsilon_0$ is an elementary kernel for every $\lambda > 0$, the elementary kernel determined by the measure $\lambda\rho_\lambda$.*

Proof. For every $\lambda' \in]0, \lambda[$ we have

$$\rho_{\lambda'} = \sum_{n=0}^{\infty} (\lambda - \lambda')^n \rho_\lambda^{n+1}. \tag{5}$$

In fact, denoting by ψ the continuous negative definite function on Γ associated with $(\mu_t)_{t>0}$, we find

$$\mathscr{F}_G\left(\sum_{n=0}^{\infty} (\lambda - \lambda')^n \rho_\lambda^{n+1}\right) = \sum_{n=0}^{\infty} (\lambda - \lambda')^n \left(\frac{1}{\lambda + \psi}\right)^{n+1} = \frac{1}{\lambda' + \psi} = \mathscr{F}_G(\rho_{\lambda'}).$$

Letting $\lambda' \searrow 0$ in (5) we get

$$\kappa = \sum_{n=0}^{\infty} \lambda^n \rho_\lambda^{n+1},$$

and hence

$$\lambda\kappa + \varepsilon_0 = \sum_{n=0}^{\infty} (\lambda\rho_\lambda)^n. \quad \square$$

13.8. Remark. The set of potential kernels on G is a cone but in general not a convex cone.

If κ is potential kernel for the transient convolution semigroup $(\mu_t)_{t>0}$, the measure $\alpha\kappa$, where $\alpha > 0$, is potential kernel for the transient convolution semigroup $(\mu_{\frac{1}{\alpha}t})_{t>0}$.

Consider the elementary kernels κ_2 and κ_3 on $G=\mathbb{R}$ defined by

$$\kappa_2=\sum_{n=0}^{\infty}\varepsilon_2^n \quad \text{and} \quad \kappa_3=\sum_{n=0}^{\infty}\varepsilon_3^n.$$

The support of the measure $\kappa_2+\kappa_3$

$$\begin{aligned}\operatorname{supp}(\kappa_2+\kappa_3)&=\operatorname{supp}(\kappa_2)\cup\operatorname{supp}(\kappa_3)\\&=\{0,2,4,\ldots\}\cup\{0,3,6,\ldots\},\end{aligned}$$

is not a semigroup under addition, and it follows by Proposition 13.3 that $\kappa_2+\kappa_3$ is not a potential kernel on $\mathbb{R}$.

For another application of the above argument see 15.24.

13.9. Example. The translation semigroup $(\varepsilon_t)_{t>0}$ on $\mathbb{R}$ is transient, and the potential kernel κ for $(\varepsilon_t)_{t>0}$ is the restriction of the Lebesgue measure to the interval $]0,\infty[$, i.e.

$$\kappa=1_{]0,\infty[}(x)\,dx. \tag{6}$$

A translation semigroup $(\varepsilon_{x(t)})_{t>0}$ on G is transient if and only if $\overline{\varphi(\mathbb{R})}$ is not compact, where $\varphi\colon \mathbb{R}\to G$ is the continuous homomorphism defining $(\varepsilon_{x(t)})_{t>0}$, cf. 8.11, and the potential kernel is in the affirmative case the image measure $\varphi(\kappa)$ of κ by φ defined by

$$\langle\varphi(\kappa),f\rangle=\int_0^{\infty}f(\varphi(x))\,dx \quad \text{for } f\in C_c(G), \tag{7}$$

where κ is given by (6), i.e. $\varphi(\kappa)$ is the restriction of a Haar measure on $\varphi(\mathbb{R})$ to the "positive half-line" $\varphi(]0,\infty[)$.

To see this, we remark that there are the following two possibilities for a continuous homomorphism $\varphi\colon \mathbb{R}\to G$, cf. Hewitt and Ross [1] p. 84:

(i) $\overline{\varphi(\mathbb{R})}$ is compact.

(ii) $\varphi\colon \mathbb{R}\to\varphi(\mathbb{R})\subseteq G$ is a homeomorphism and $\varphi(\mathbb{R})$ is a closed subgroup of G.

In the first case $(\varepsilon_{x(t)})_{t>0}$ is clearly recurrent, because all the measures $\varepsilon_{x(t)}$ are concentrated on the compact set $\overline{\varphi(\mathbb{R})}$, cf. Remark 13.4.

In the second case $(\varepsilon_{x(t)})_{t>0}$ is transient, because for every $f\in C_c^+(G)$ the set $\operatorname{supp}(f)\cap\varphi(\mathbb{R})$ is compact, and there exists consequently a $t_0>0$ such that

$$\langle\varepsilon_{x(t)},f\rangle=f(\varphi(t))=0 \quad \text{for } t\geqq t_0.$$

It is clear that the potential kernel for $(\varepsilon_{x(t)})_{t>0}$ in the transient case is given by (7).

13.10. Proposition. *Let $(\mu_t)_{t>0}$ be a transient convolution semigroup on G with potential kernel κ. The measure $\frac{1}{2}(\kappa+\check{\kappa})$ is positive definite and κ is shift-bounded.*

Proof. Let $(\rho_\lambda)_{\lambda>0}$ be the resolvent for $(\mu_t)_{t>0}$. The measure $\frac{1}{2}(\rho_\lambda+\check{\rho}_\lambda)$ is positive definite for $\lambda>0$; to see this we calculate the Fourier transform of $\frac{1}{2}(\rho_\lambda+\check{\rho}_\lambda)$, cf. 8.12,

$$\mathscr{F}_G[\tfrac{1}{2}(\rho_\lambda+\check{\rho}_\lambda)](\gamma)=\operatorname{Re}\frac{1}{\lambda+\psi(\gamma)}=\frac{\lambda+\operatorname{Re}\psi(\gamma)}{|\lambda+\psi(\gamma)|^2} \quad \text{for } \gamma\in\Gamma, \tag{8}$$

which is a non-negative function on Γ, and the assertion follows from Proposition 4.14.

It follows that $\frac{1}{2}(\kappa+\check{\kappa})$ is positive definite, since it is vague limit of the measures $\frac{1}{2}(\rho_\lambda+\check{\rho}_\lambda)$ as $\lambda\searrow 0$.

In particular $\frac{1}{2}(\kappa+\check{\kappa})$ is shift-bounded (cf. Proposition 4.4), and the measure κ is therefore also shift-bounded. □

The Fourier transform of the positive definite measure $\frac{1}{2}(\kappa+\check{\kappa})$ will be determined in Theorem 13.33.

13.11. Remark. The notion of positive definiteness can be weakened in the following way: A measure μ on G is said to have property (P) if

$$\langle\mu, f*\check{f}\rangle\geqq 0 \quad \text{for all real functions } f\in C_c(G).$$

It is easy to verify the following statements.

(i) The set of measures having property (P) is a vaguely closed convex cone in $M(G)$ which contains the cone of positive definite measures and which is stable under reflection and conjugation; therefore $\check{\mu}$, $\bar{\mu}$ and $\tilde{\mu}$ have property (P) whenever μ has property (P).

(ii) A measure μ on G has property (P) if and only if $\mu+\check{\mu}$ is positive definite. In particular a symmetric measure has property (P) if and only if it is positive definite.

(iii) A positive (or real) bounded measure μ on G has property (P) if and only if $\operatorname{Re}\hat{\mu}(\gamma)\geqq 0$ for all $\gamma\in\Gamma$.

(iv) A positive measure having property (P) is shift-bounded.

The assertion of Proposition 13.10 may then be expressed in the following way: the potential kernel κ has property (P) and κ is consequently shift-bounded. In fact, since $\operatorname{Re}\hat{\rho}_\lambda\geqq 0$ for $\lambda>0$, cf. (8), it follows by (iii) that ρ_λ has property (P) for $\lambda>0$, and this implies by (i) that κ has property (P).

13.12. Exercise. The periodicity group $\operatorname{per}(\kappa)$ for a potential kernel κ on G is equal to $\{0\}$.

Many statements in the sequel involve the functions $\frac{1}{\psi}$ and $\operatorname{Re}\frac{1}{\psi}$, where ψ is a continuous negative definite function on Γ. An expression as "$\frac{1}{\psi}$ is integrable over V" will be used as an abbreviation for "$\psi\neq 0$ a.e. in V, and the function $\frac{1}{\psi}$, which is defined a.e. in V, is integrable over V". Likewise an expression as "$\operatorname{Re}\frac{1}{\psi}$

is locally integrable on Γ" is an abbreviation for "$\psi \neq 0$ locally a.e. in Γ, and the function $\operatorname{Re}\frac{1}{\psi}$, which is defined locally a.e. in Γ, is locally integrable".

The next result can be viewed as a generalization of Corollary 7.9.

13.13. Proposition. *Let ψ be a continuous negative definite function on Γ, and suppose that $\frac{1}{\psi}$ is integrable over some open, relatively compact neighbourhood of 0. Then $\frac{1}{\psi}$ is locally integrable, and the measure $\frac{1}{\psi}\omega_\Gamma$ is positive definite.*

Proof. If $\psi(0)>0$ the assertion follows immediately from Corollary 7.9 and Proposition 4.1, and for the rest of the proof we suppose that $\psi(0)=0$.

Let $K \subseteq \Gamma$ be compact. If $\psi(\gamma) \neq 0$ for all $\gamma \in K$, then

$$\int_K \frac{1}{|\psi(\gamma)|}\, d\gamma < \infty .$$

If $\psi(\gamma)=0$ for some $\gamma \in K$, the set $K_0=\{\gamma \in K \mid \psi(\gamma)=0\}$ is a compact subset of Γ and we may therefore choose $\gamma_1, \ldots, \gamma_n \in K_0$ such that $K_0 \subseteq \bigcup_{i=1}^n (\gamma_i + V)$, where V is an open, relatively compact neighbourhood of 0 such that $\frac{1}{\psi}$ is integrable over V. The set

$$F = K \setminus \bigcup_{i=1}^{n} (\gamma_i + V)$$

is a compact subset of K with the property that $\psi(\gamma) \neq 0$ for all $\gamma \in F$. It follows that $\psi \neq 0$ almost everywhere in K and

$$\begin{aligned}\int_K \frac{1}{|\psi(\gamma)|}\, d\gamma &\leqq \int_F \frac{1}{|\psi(\gamma)|}\, d\gamma + \int_{\bigcup_{i=1}^n (\gamma_i+V)} \frac{1}{|\psi(\gamma)|}\, d\gamma \\ &\leqq \int_F \frac{1}{|\psi(\gamma)|}\, d\gamma + n \int_V \frac{1}{|\psi(\gamma)|}\, d\gamma < \infty,\end{aligned}$$

where we have used that $\psi(\gamma_0+\gamma)=\psi(\gamma)$ for $\gamma \in \Gamma$ and $\gamma_0 \in K_0$ since $K_0 \subseteq \operatorname{per}(\psi)$, cf. Proposition 8.27.

It remains to see that the measure $\frac{1}{\psi}\omega_\Gamma$ is positive definite. Let $f \in C_c(\Gamma)$. For every $n \in \mathbb{N}$ we have

$$\int f * \tilde{f}(\gamma) \frac{1}{\psi(\gamma)+\frac{1}{n}}\, d\gamma \geqq 0,$$

because $\left(\psi+\frac{1}{n}\right)^{-1}$ is continuous and positive definite, cf. 7.9 and 4.1. The function $\gamma\mapsto f*\tilde{f}(\gamma)\frac{1}{\psi(\gamma)}$ is defined a.e. in Γ and

$$\lim_{n\to\infty} f*\tilde{f}(\gamma)\frac{1}{\psi(\gamma)+\frac{1}{n}}=f*\tilde{f}(\gamma)\frac{1}{\psi(\gamma)} \qquad \text{a.e. in } \Gamma.$$

Since $\operatorname{Re}\psi\geqq 0$, we have $|\psi|\leqq\left|\psi+\frac{1}{n}\right|$ for all $n\in\mathbb{N}$, and the dominated convergence theorem gives that

$$\int f*\tilde{f}(\gamma)\frac{1}{\psi(\gamma)}d\gamma=\lim_{n\to\infty}\int f*\tilde{f}(\gamma)\frac{1}{\psi(\gamma)+\frac{1}{n}}d\gamma\geqq 0. \quad \square$$

13.14. Exercise. Let ψ be a continuous negative definite function on Γ and suppose that $\operatorname{Re}\frac{1}{\psi}$ is integrable over some open, relatively compact neighbourhood of 0. Then $\operatorname{Re}\frac{1}{\psi}$ is locally integrable on Γ and the measure $\operatorname{Re}\frac{1}{\psi}\omega_\Gamma$ is positive.

13.15. Proposition. *Let $(\mu_t)_{t>0}$ be a transient convolution semigroup on G with associated continuous negative definite function ψ on Γ. The function $\operatorname{Re}\frac{1}{\psi}$ is locally integrable on Γ.*

Proof. We may assume that $\psi(0)=0$. The set $\Gamma_0=\{\gamma\in\Gamma\mid\psi(\gamma)=0\}$ is a closed subgroup of Γ and $\operatorname{supp}(\mu_t)\subseteq\Gamma_0^\perp$ for $t>0$ by Proposition 8.27. Since $(\mu_t)_{t>0}$ is transient, $\Gamma_0^\perp$ is not compact (cf. 13.4) and therefore Γ_0 is not open. A closed subgroup of Γ is either open or locally negligible (cf. Bourbaki [3] p. 169), and it follows that $\psi\neq 0$ locally a.e. on Γ.

Let $K\subseteq\Gamma$ be compact and choose $f\in C_c^+(G)$ such that $|\bar{\mathscr{F}}_G f|\geqq 1$ on K (cf. Proposition 2.4). By Fatou's lemma we then find

$$\begin{aligned}\int_K \operatorname{Re}\frac{1}{\psi(\gamma)}d\gamma &= \int_K \lim_{\lambda\to 0}\operatorname{Re}\frac{1}{\lambda+\psi(\gamma)}d\gamma\\ &\leqq \int_\Gamma \liminf_{\lambda\to 0}|\bar{\mathscr{F}}_G f(\gamma)|^2\operatorname{Re}\frac{1}{\lambda+\psi(\gamma)}d\gamma\\ &\leqq \liminf_{\lambda\to 0}\int_\Gamma |\bar{\mathscr{F}}_G f(\gamma)|^2\operatorname{Re}\frac{1}{\lambda+\psi(\gamma)}d\gamma\\ &= \liminf_{\lambda\to 0}\frac{1}{2}\langle\rho_\lambda+\check{\rho}_\lambda, f*\tilde{f}\rangle\\ &= \frac{1}{2}\langle\kappa+\check{\kappa}, f*\tilde{f}\rangle<\infty,\end{aligned}$$

where $(\rho_\lambda)_{\lambda>0}$ is the resolvent for $(\mu_t)_{t>0}$ and κ is the potential kernel for $(\mu_t)_{t>0}$, and the assertion follows. □

13.16. Corollary. *Let μ be a positive bounded measure on G with total mass $\mu(G)\leqq 1$. If the series $\sum_{n=0}^{\infty}\mu^n$ converges vaguely then the function $\operatorname{Re}\frac{1}{1-\hat{\mu}}$ is locally integrable on Γ.*

Proof. This is immediate by Proposition 13.15, since the continuous negative definite function on Γ associated with the transient (cf. 13.5) convolution semigroup $(e^{-t}\exp(t\mu))_{t>0}$ on G, determined by μ, is the function $1-\hat{\mu}$. □

13.17. Remark. The assertion of Proposition 13.15 can be stated in the form of the implication

$$(\mu_t)_{t>0} \text{ transient } \Rightarrow \operatorname{Re}\frac{1}{\psi} \text{ locally integrable.} \tag{9}$$

The converse implication

$$\operatorname{Re}\frac{1}{\psi} \text{ locally integrable } \Rightarrow (\mu_t)_{t>0} \text{ transient,} \tag{10}$$

is also true. When G is compact this follows easily from 13.4, and the special case where $(\mu_t)_{t>0}$ is symmetric is treated in 13.27. See also 13.28. However, probabilistic methods are deeply involved in the arguments for the general case which will thus not be discussed here, and it seems to be difficult to find a purely analytic proof, cf. 13.42.

In the proof of (10) it is no restriction to assume that the smallest closed subgroup G_0 of G which contains $\operatorname{supp}(\mu_t)$ for all $t>0$ is equal to G. In fact, it is easy to see that $\operatorname{Re}\frac{1}{\psi}$ is locally integrable on Γ if and only if $\operatorname{Re}\frac{1}{\dot{\psi}}$ is locally integrable on Γ/Γ_0 where $\dot{\psi}$ is the quotient function of ψ over the periodicity group $\Gamma_0=G_0^{\perp}$ for ψ, and $\dot{\psi}$ is the continuous negative definite function associated with $(\mu_t)_{t>0}$ considered as a convolution semigroup on G_0, cf. 8.28.

Furthermore it is enough to prove (10) for convolution semigroups determined by a probability measure. For a convolution semigroup $(e^{-t}\exp(t\mu))_{t>0}$ determined by a probability measure μ on G the implication (10) takes the form (cf. Lemma 13.5)

$$\operatorname{Re}\frac{1}{1-\hat{\mu}} \text{ locally integrable } \Rightarrow \sum_{n=0}^{\infty}\mu^n \text{ converges vaguely,} \tag{11}$$

which is the converse of the statement in Corollary 13.16.

In order to see that (10) is a consequence of (11) we consider a convolution semigroup $(\mu_t)_{t>0}$ on G such that $\operatorname{Re}\frac{1}{\psi}$ is locally integrable on Γ. Denoting by

ρ_1 the resolvent measure corresponding to $\lambda=1$ we find that the function

$$\operatorname{Re}\frac{1}{1-\hat{\rho}_1}=1+\operatorname{Re}\frac{1}{\psi}$$

is locally integrable on Γ, and it follows by (11) that the series $\sum_{n=0}^{\infty}\rho_1^n$ converges vaguely. For $\lambda\in\left]0,\frac{1}{2}\right[$ we have (cf. (5))

$$\rho_\lambda=\sum_{n=0}^{\infty}(1-\lambda)^n\rho_1^{n+1}=\frac{1}{1-\lambda}\sum_{n=1}^{\infty}(1-\lambda)^n\rho_1^n\leqq 2\sum_{n=0}^{\infty}\rho_1^n,$$

and it follows that $(\mu_t)_{t>0}$ is transient.

Thus (10) follows when (11) is established for probability measures μ on G such that the smallest closed subgroup of G containing the support of μ is G, and this is done in Port and Stone [1] p. 46.

The implication (11) was first established for $G=\mathbb{Z}^n$ by Spitzer [1] and for $G=\mathbb{R}^n$ by Ornstein [1].

13.18. A convolution semigroup $(\mu_t)_{t>0}$ on G induces a translation invariant Feller semigroup $(P_t)_{t>0}$ on G, cf. 12.9. The resolvent, the potential operator and the zero-resolvent for $(P_t)_{t>0}$ are denoted respectively $(N_\lambda)_{\lambda>0}$, $(N, D(N))$ and $(N_0, D(N_0))$.

If $(\mu_t)_{t>0}$ is transient, then $(N_0, D(N_0))$ is densely defined. To see this we remark that by Proposition 13.15 we have $\psi\neq 0$ locally a.e. on Γ, where ψ is the associated continuous negative definite function on Γ, and Exercise 12.13 now gives that $D(N_0)$ is dense in $C_0(G)$.

The potential operator $(N, D(N))$ need not be densely defined in the transient case as the example of the translation semigroup $(\varepsilon_t)_{t>0}$ on $\mathbb{R}$ shows, cf. 12.15 and 13.9. On the other hand $D(N)$ (and hence also $D(N_0)$) can be dense in $C_0(G)$ for recurrent semigroups, e.g. for the Brownian semigroup in $\mathbb{R}^n$ for $n=1,2$, cf. 12.25 and 13.29.

There is thus no obvious relation between the two "strengthenings" of the condition that $D(N_0)$ is dense: transience and density of $D(N)$, and we shall now discuss a condition on $(\mu_t)_{t>0}$ which is "stronger" than transience of $(\mu_t)_{t>0}$ (cf. 13.21) and implies the density of $D(N)$. We start with the following lemma.

13.19. Lemma. *The two conditions*

(i) $C_c(G)\subseteq D(N)$,

(ii) $C_c(G)\subseteq D(N_0)$,

are equivalent.

Proof. Since $D(N)\subseteq D(N_0)$, cf. 11.15, it is clear that (i) implies (ii). Suppose conversely that $C_c(G)\subseteq D(N_0)$ and consider $f\in C_c^+(G)$. The function $N_0 f$ belongs to $C_0^+(G)$, and for every $x\in G$ we have

$$N_0 f(x)=\lim_{\lambda\to 0}N_\lambda f(x)=\int_0^{\infty}P_s f(x)\,ds$$

by the monotone convergence theorem, from which it also follows that

$$\lim_{t\to\infty} \int_0^t P_s f(x)\,ds = N_0 f(x) \quad \text{for } x\in G. \tag{12}$$

This however implies that

$$\lim_{t\to\infty} \int_0^t P_s f\,ds = N_0 f \quad \text{uniformly on } G.$$

In fact, for a given $\varepsilon>0$ there exists a compact set $K\subseteq G$ such that

$$0\leqq N_0 f(x)\leqq \varepsilon \quad \text{for all } x\notin K.$$

The convergence in (12) is increasing, and it follows by Dini's theorem that the convergence in (12) is uniform on K, i.e. there exists a $t_0>0$ such that

$$\left|\int_0^t P_s f(x)\,ds - N_0 f(x)\right| \leqq \varepsilon \quad \text{for } t\geqq t_0 \text{ and } x\in K.$$

This shows that

$$\lim_{t\to\infty}\left\|\int_0^t P_s f\,ds - N_0 f\right\|_\infty = 0,$$

and it follows that $f\in D(N)$, which implies that $C_c(G)\subseteq D(N)$. □

13.20. Definition. A convolution semigroup $(\mu_t)_{t>0}$ on G is said to be *integrable* if the two equivalent conditions of Lemma 13.19 are fulfilled.

For the Feller semigroup induced by an integrable convolution semigroup on G, it follows from 11.9 and 11.15 that $N=N_0(=-A^{-1})$ is densely defined on $C_0(G)$.

There is the following relation between the notion of transience and the notion of integrability.

13.21. Proposition. *A convolution semigroup $(\mu_t)_{t>0}$ on G is integrable if and only if $(\mu_t)_{t>0}$ is transient and the potential kernel κ for $(\mu_t)_{t>0}$ tends to zero at infinity.*

Proof. Suppose first that $(\mu_t)_{t>0}$ is integrable. The zero-resolvent $(N_0, D(N_0))$ for $(P_t)_{t>0}$ is translation invariant, cf. Exercise 12.11, and the restriction of N_0 to $C_c(G)$ defines therefore a translation invariant continuous kernel on G. It follows by Lemma 12.2 that there exists a positive measure κ on G such that

$$N_0 f(x) = \kappa * f(x) \quad \text{for } f\in C_c(G) \text{ and } x\in G.$$

It is clear that κ tends to zero at infinity. For $f \in C_c^+(G)$ we find

$$\lim_{\lambda \to 0} \langle \rho_\lambda, f \rangle = \lim_{\lambda \to 0} N_\lambda(\check{f})(0) = N_0(\check{f})(0) = \kappa * \check{f}(0) = \langle \kappa, f \rangle,$$

where $(\rho_\lambda)_{\lambda>0}$ is the resolvent for $(\mu_t)_{t>0}$, and this shows that $(\mu_t)_{t>0}$ is transient and that κ is the potential kernel for $(\mu_t)_{t>0}$.

Suppose now conversely that $(\mu_t)_{t>0}$ is transient and that the potential kernel κ for $(\mu_t)_{t>0}$ tends to zero at infinity. For $f \in C_c^+(G)$ we clearly have

$$\lim_{\lambda \to 0} N_\lambda f(x) = \lim_{\lambda \to 0} \langle \rho_\lambda, \tau_x(\check{f}) \rangle = \langle \kappa, \tau_x(\check{f}) \rangle = \kappa * f(x)$$

pointwise and increasingly on G. Since $\kappa * f$ belongs to $C_0(G)$ this implies as in 13.19 that

$$\lim_{\lambda \to 0} N_\lambda f = \kappa * f \quad \text{uniformly on } G,$$

i.e. $f \in D(N_0)$ and $N_0 f = \kappa * f$. This shows that $C_c^+(G) \subseteq D(N_0)$ and $(\mu_t)_{t>0}$ is consequently integrable. □

13.22. Remark. From the proof of Proposition 13.21 we get that the zero-resolvent $(N_0, D(N_0))$ for $(P_t)_{t>0}$ in the integrable case is given by

$$N_0 f(x) = \kappa * f(x) \qquad \text{for } x \in G,$$

for all $f \in C_c(G)$ and more generally for $f \in D(N_0)^+$.

13.23. Proposition (Berg [1]). *Let $(\mu_t)_{t>0}$ be a convolution semigroup on G with associated continuous negative definite function ψ on Γ. If the function $1/\psi$ is locally integrable, then $(\mu_t)_{t>0}$ is integrable and the potential kernel κ for $(\mu_t)_{t>0}$ is given by*

$$\kappa = \mathscr{F}_\Gamma\left[\left(\frac{1}{\psi}\omega_\Gamma\right)^{\vee}\right].$$

Proof. The measure $\left(\frac{1}{\psi}\omega_\Gamma\right)^{\vee} = \frac{1}{\bar{\psi}}\omega_\Gamma$ is by Proposition 13.13 positive definite, and the Fourier transform

$$\sigma = \mathscr{F}_\Gamma\left(\frac{1}{\bar{\psi}}\omega_\Gamma\right)$$

is a positive measure on G, which tends to zero at infinity (cf. Proposition 4.9). For $\varphi \in C_c(\Gamma)$ we have

$$\int_G |\bar{\mathscr{F}}_\Gamma \varphi(x)|^2 d\sigma(x) = \int_\Gamma \varphi * \tilde{\varphi}(\gamma) \frac{1}{\bar{\psi}(\gamma)} d\gamma \tag{13}$$

and

$$\int_G |\bar{\mathscr{F}}_\Gamma \varphi(x)|^2 d\rho_\lambda(x) = \int_\Gamma \varphi * \tilde{\varphi}(\gamma) \frac{1}{\lambda + \bar{\psi}(\gamma)} d\gamma \quad \text{for } \lambda > 0,$$

where $(\rho_\lambda)_{\lambda>0}$ is the resolvent for $(\mu_t)_{t>0}$, and as in the proof of Proposition 13.13, we get by the dominated convergence theorem that

$$\lim_{\lambda\to 0} \int_\Gamma \varphi * \tilde{\varphi}(\gamma) \frac{1}{\lambda + \bar{\psi}(\gamma)} d\gamma = \int_\Gamma \varphi * \tilde{\varphi}(\gamma) \frac{1}{\bar{\psi}(\gamma)} d\gamma,$$

i.e.

$$\lim_{\lambda\to 0} \int_G |\bar{\mathscr{F}}_\Gamma \varphi(x)|^2 d\rho_\lambda(x) = \int_G |\bar{\mathscr{F}}_\Gamma \varphi(x)|^2 d\sigma(x) < \infty. \tag{14}$$

For every $f \in C_c^+(G)$ there exists $\varphi \in C_c(\Gamma)$ such that $f \leq |\bar{\mathscr{F}}_\Gamma \varphi|^2$, cf. Proposition 2.4, and it follows that

$$\lim_{\lambda\to 0} \langle \rho_\lambda, f \rangle \leq \lim_{\lambda\to 0} \int_G |\bar{\mathscr{F}}_\Gamma \varphi(x)|^2 d\rho_\lambda(x) < \infty,$$

which shows that $(\mu_t)_{t>0}$ is transient. Let κ be the potential kernel for $(\mu_t)_{t>0}$. By (2) we have

$$\int_G |\bar{\mathscr{F}}_\Gamma \varphi(x)|^2 d\kappa(x) = \lim_{\lambda\to 0} \int_G |\bar{\mathscr{F}}_\Gamma \varphi(x)|^2 d\rho_\lambda(x) \quad \text{for } \varphi \in C_c(\Gamma),$$

so by (13) and (14) we have for $\varphi \in C_c(\Gamma)$

$$\int_G |\bar{\mathscr{F}}_\Gamma \varphi(x)|^2 d\kappa(x) = \int_G |\bar{\mathscr{F}}_\Gamma \varphi(x)|^2 d\sigma(x) = \int_\Gamma \varphi * \tilde{\varphi}(\gamma) \frac{1}{\bar{\psi}(\gamma)} d\gamma,$$

and it follows by Theorem 4.7 that $\kappa = \sigma$. Finally $(\mu_t)_{t>0}$ is integrable since $\kappa = \sigma$ tends to zero at infinity, cf. Proposition 13.21. □

13.24. Remark. With the obvious definition of the co-Fourier transform $\bar{\mathscr{F}}_\Gamma \mu$ of a positive definite measure μ on Γ as

$$\bar{\mathscr{F}}_\Gamma \mu = \mathscr{F}_\Gamma(\check{\mu}),$$

the second part of Proposition 13.23 can be expressed that κ is the co-Fourier transform of the positive definite measure $\frac{1}{\psi}\omega_\Gamma$.

13.25. Remark. The first part of Proposition 13.23 can be stated as the implication

$$\frac{1}{\psi} \text{ locally integrable} \Rightarrow (\mu_t)_{t>0} \text{ integrable}.$$

The converse implication

$$(\mu_t)_{t>0} \text{ integrable } \Rightarrow \frac{1}{\psi} \text{ locally integrable}$$

holds for symmetric and "almost-symmetric" semigroups, cf. Corollary 13.27 and Exercise 13.28, but it is not true in general as shown by the following example.

13.26. Example (Berg [1]). Let $G=\mathbb{Z}$ and let $a>0$ be determined such that

$$\mu=\sum_{n=2}^{\infty}\frac{a}{n^2\log n}\varepsilon_n$$

is a probability measure. The elementary kernel

$$\kappa=\sum_{n=0}^{\infty}\mu^n$$

determined by μ exists because the support of μ^n is contained in the set $\{2n, 2n+1, \ldots\}$, $n\in\mathbb{N}$. We have that $\kappa(n)=0$ for $n<0$.

The first moment of μ is infinite and it follows by a result in Spitzer [1] p. 100 that

$$\lim_{n\to\infty}\kappa(n)=0$$

i.e. κ tends to zero at infinity.

The convolution semigroup $(\mu_t)_{t>0}$ determined by μ is consequently integrable and the associated negative definite function is

$$\psi(\theta)=1-\hat{\mu}(\theta)=\sum_{n=2}^{\infty}\frac{a}{n^2\log n}(1-\cos(n\theta))+i\sum_{n=2}^{\infty}\frac{a}{n^2\log n}\sin(n\theta)$$

for $\theta\in\mathbb{R}$, when identifying functions on $\hat{\mathbb{Z}}$ with functions on $\mathbb{R}$ which are periodic with period 2π.

From Zygmund [1] p. 229 we find

$$\psi(\theta)\sim -ia\theta\log|\log\theta| \quad \text{for } \theta\to 0^+,$$

which shows that the function $1/\psi$ is not integrable over a neighbourhood of 0.

13.27. Corollary. *Let $(\mu_t)_{t>0}$ be a symmetric convolution semigroup on G with associated continuous negative definite function ψ on Γ. The following three conditions are equivalent:*

(i) *$(\mu_t)_{t>0}$ is integrable.*

(ii) *$(\mu_t)_{t>0}$ is transient.*

(iii) *$1/\psi$ is locally integrable on Γ.*

Proof. The result is contained in the Propositions 13.15, 13.21 and 13.23 since ψ is real. □

13.28. Exercise. Let $(\mu_t)_{t>0}$ be a convolution semigroup on G with associated continuous negative definite function ψ on Γ, and suppose that ψ satisfies an inequality of the form

$$|\operatorname{Im}\psi| \leqq C \operatorname{Re}\psi \quad \text{on } \Gamma$$

(or just in some neighbourhood of 0) for some constant $C>0$. The following conditions are equivalent:

(i) $(\mu_t)_{t>0}$ is integrable.
(ii) $(\mu_t)_{t>0}$ is transient.
(iii) $1/\psi$ is locally integrable on Γ.

13.29. Example. The Brownian semigroup $(\mu_t)_{t>0}$ on $\mathbb{R}^n$, cf. 10.3, is recurrent for $n=1,2$ and integrable for $n \geqq 3$. In fact, the associated continuous negative definite function on $\mathbb{R}^n$ is $y \mapsto \|y\|^2$, and it is easily seen that the function $y \mapsto \|y\|^{-2}$ is locally integrable on $\mathbb{R}^n$ if and only if $n \geqq 3$, and the assertion follows by Corollary 13.27.

This may also be seen directly. The measure μ_t has the function

$$g_t(x) = (4\pi t)^{-n/2} \exp\left(-\frac{\|x\|^2}{4t}\right) \quad \text{for } x \in \mathbb{R}^n,$$

as density with respect to the Lebesgue measure on $\mathbb{R}^n$, and a simple calculation gives that for $n=1,2$

$$\int_0^\infty g_t(x)\,dt = \infty \quad \text{for all } x \in \mathbb{R}^n,$$

while for $n \geqq 3$

$$\int_0^\infty g_t(x)\,dt = \begin{cases} \infty & \text{for } x=0, \\ \dfrac{\Gamma\left(\frac{n}{2}\right)}{2\pi^{n/2}(n-2)} \|x\|^{2-n} & \text{for } x \neq 0. \end{cases}$$

For $n \geqq 3$ we put

$$N_n(x) = \begin{cases} \infty & \text{for } x=0, \\ \|x\|^{2-n} & \text{for } x \neq 0, \end{cases} \tag{15}$$

and

$$k_n = \frac{\Gamma\left(\frac{n}{2}\right)}{2\pi^{n/2}(n-2)}.$$

The function (15) is locally integrable on $\mathbb{R}^n$, and by Fubini's theorem we find that $(\mu_t)_{t>0}$ is recurrent in the case $n=1,2$, while $(\mu_t)_{t>0}$ is transient in the case

$n \geqq 3$ with potential kernel

$$\kappa_n = k_n N_n(x)\,dx.$$

The measure κ_n, which tends to zero at infinity, is (a multiple of) the *Newtonian kernel* on $\mathbb{R}^n$.

13.30. Example. The heat semigroup $(\mu_t)_{t>0}$ on $\mathbb{R}^n \times \mathbb{R}$, (cf. 10.4), is integrable for all $n \geqq 1$. The resolvent $(\rho_\lambda)_{\lambda>0}$ for $(\mu_t)_{t>0}$ is given by

$$\begin{aligned}\langle \rho_\lambda, g\rangle &= \int_0^\infty e^{-\lambda s}\langle \mu_s, g\rangle\,ds\\ &= \int_0^\infty e^{-\lambda s}\left[\int_{\mathbb{R}^n} g(x,s)(4\pi s)^{-n/2}\exp\left(-\frac{\|x\|^2}{4s}\right)dx\right]ds\end{aligned}$$

for $\lambda>0$ and $g(\cdot,\cdot)\in C_c(\mathbb{R}^n\times\mathbb{R})$, and since the function

$$k(x,s)=\begin{cases}(4\pi s)^{-n/2}\exp\left(-\dfrac{\|x\|^2}{4s}\right) & \text{for } x\in\mathbb{R}^n \text{ and } s>0,\\ 0 & \text{for } x\in\mathbb{R}^n \text{ and } s\leqq 0,\end{cases}$$

is locally integrable, it follows that $(\mu_t)_{t>0}$ is transient and that the potential kernel κ for $(\mu_t)_{t>0}$ has $k(\cdot,\cdot)$ as density with respect to the Lebesgue measure on $\mathbb{R}^n\times\mathbb{R}$. It is easy to see that the function $k(\cdot,\cdot)$, which is continuous in $(\mathbb{R}^n\times\mathbb{R})\setminus\{(0,0)\}$, tends to zero at infinity, and it follows that κ tends to zero at infinity, and $(\mu_t)_{t>0}$ is consequently integrable, cf. Proposition 13.21.

The continuous negative definite function on $\mathbb{R}^n\times\mathbb{R}$ associated with $(\mu_t)_{t>0}$ is given by

$$\psi(y,u)=\|y\|^2+iu \qquad \text{for } (y,u)\in\mathbb{R}^n\times\mathbb{R},$$

and the function $1/\psi$ is locally integrable on $\mathbb{R}^n\times\mathbb{R}$, so the above assertion is also a consequence of Proposition 13.23.

13.31. Exercise. Let $(\mu_t)_{t>0}$ and $(\nu_t)_{t>0}$ be convolution semigroups on G and suppose that $(\mu_t)_{t>0}$ is symmetric and transient (or integrable). Then the "mixed" convolution semigroup $(\mu_t * \nu_t)_{t>0}$ on G is integrable. For transient convolution semigroups $(\mu_t)_{t>0}$ and $(\nu_t)_{t>0}$ on G the convolution semigroup $(\mu_t * \nu_t)_{t>0}$ need not be transient.

13.32. Exercise. Let $(\mu_t)_{t>0}$ (resp. $(\nu_t)_{t>0}$) be a convolution semigroup on the LCA-group G_1 (resp. G_2) and suppose that $(\mu_t)_{t>0}$ is symmetric and integrable. The product convolution semigroup $(\mu_t \otimes \nu_t)_{t>0}$ on $G_1\times G_2$ is integrable.

13.33. Theorem (Berg [1]). *Let $(\mu_t)_{t>0}$ be a transient convolution semigroup on G with associated continuous negative definite function ψ on Γ, and let κ be the* *potential kernel for* $(\mu_t)_{t>0}$. *The Fourier transform of the measure* $\frac{1}{2}(\kappa+\check{\kappa})$ *is of the form*

$$\mathscr{F}_G\left(\tfrac{1}{2}(\kappa+\check{\kappa})\right)=\omega+\operatorname{Re}\frac{1}{\psi}\,\omega_\Gamma, \tag{16}$$

where ω *is a translation invariant positive measure on the closed subgroup* $\Gamma_0=\{\gamma\in\Gamma\mid\psi(\gamma)=0\}$ *of* Γ (*in particular* $\omega=0$ *if* $\psi(0)>0$), *possibly the zero measure.*

The potential kernel κ *tends to zero at infinity if and only if* $\omega=0$.

Proof. Suppose first that $\psi(0)>0$. The potential kernel κ is then a bounded measure, whose Fourier transform is given by

$$\mathscr{F}_G\,\kappa(\gamma)=\frac{1}{\psi(\gamma)}\qquad\text{for } \gamma\in\Gamma,$$

cf. 13.4.

The formula (16) follows ($\omega=0$), and κ tends to zero at infinity because κ is a bounded measure.

For the rest of the proof we suppose $\psi(0)=0$.

Putting $\nu=\frac{1}{2}(\kappa+\check{\kappa})$ we have that

$$\nu=\lim_{\lambda\to 0}\tfrac{1}{2}(\rho_\lambda+\check{\rho}_\lambda)\quad\text{vaguely},$$

where $(\rho_\lambda)_{\lambda>0}$ is the resolvent for $(\mu_t)_{t>0}$ and by Theorem 4.16 this implies

$$\mathscr{F}_G\,\nu=\lim_{\lambda\to 0}\mathscr{F}_G\left(\tfrac{1}{2}(\rho_\lambda+\check{\rho}_\lambda)\right)=\lim_{\lambda\to 0}\left[\operatorname{Re}\frac{1}{\lambda+\psi}\,\omega_\Gamma\right]\quad\text{vaguely}. \tag{17}$$

The smallest closed subgroup G_0 of G containing $\bigcup_{t>0}\operatorname{supp}(\mu_t)$ is σ-compact (Theorem 8.24), and it follows (cf. Bourbaki [4] p. 152) that the dual group $\hat{G}_0$ is metrizable. However $\hat{G}_0$ is isomorphic with the quotient Γ/Γ_0 (cf. 2.11 and Proposition 8.27), and it follows that Γ/Γ_0 is metrizable. Let $(U_n)_{n\in\mathbb{N}}$ be a decreasing sequence of open relatively compact neighbourhoods of 0 in Γ/Γ_0 such that $\bigcap_{n\in\mathbb{N}}U_n=\{0\}$, and let $\pi\colon\Gamma\to\Gamma/\Gamma_0$ be the canonical mapping.

For $n\in\mathbb{N}$, $\lambda>0$ and $f\in C_c(\Gamma)$ we find

$$\int_\Gamma f(\gamma)\operatorname{Re}\frac{1}{\lambda+\psi(\gamma)}\,d\gamma=\int_{\pi^{-1}(U_n)}f(\gamma)\operatorname{Re}\frac{1}{\lambda+\psi(\gamma)}\,d\gamma+\int_{\Gamma\setminus\pi^{-1}(U_n)}f(\gamma)\operatorname{Re}\frac{1}{\lambda+\psi(\gamma)}\,d\gamma. \tag{18}$$

For λ tending to zero, the left-hand side converges to $\langle\mathscr{F}_G\,\nu,f\rangle$, while the second member on the right-hand side converges to

$$\int_{\Gamma\setminus\pi^{-1}(U_n)}f(\gamma)\operatorname{Re}\frac{1}{\psi(\gamma)}\,d\gamma,$$

because $\Gamma\setminus\pi^{-1}(U_n)$ is a closed subset of Γ disjoint with Γ_0. It follows that the first member on the right-hand side converges. Denoting the limit

$$\omega_n(f)=\lim_{\lambda\to 0}\int_{\pi^{-1}(U_n)} f(\gamma)\operatorname{Re}\frac{1}{\lambda+\psi(\gamma)}\,d\gamma,$$

it is clear that the mapping $f\mapsto\omega_n(f)$ is a positive measure (denoted ω_n) on Γ, and furthermore that $(\omega_n)_{n\in\mathbb{N}}$ is a decreasing, hence a vaguely convergent sequence of positive measures on Γ. Let ω be the limit of $(\omega_n)_{n\in\mathbb{N}}$. For $n\in\mathbb{N}$ and $f\in C_c^+(\Gamma)$ we have

$$\langle\mathscr{F}_G\nu,f\rangle=\langle\omega_n,f\rangle+\int_{\Gamma\setminus\pi^{-1}(U_n)} f(\gamma)\operatorname{Re}\frac{1}{\psi(\gamma)}\,d\gamma, \tag{19}$$

and since Γ_0 is locally negligible and $\operatorname{Re}\frac{1}{\psi}$ is locally integrable on Γ, the monotone convergence theorem gives that

$$\lim_{n\to\infty}\int_{\Gamma\setminus\pi^{-1}(U_n)} f(\gamma)\operatorname{Re}\frac{1}{\psi(\gamma)}\,d\gamma=\int_{\Gamma\setminus\Gamma_0} f(\gamma)\operatorname{Re}\frac{1}{\psi(\gamma)}\,d\gamma=\int_\Gamma f(\gamma)\operatorname{Re}\frac{1}{\psi(\gamma)}\,d\gamma,$$

or by (19) that

$$\langle\mathscr{F}_G\nu,f\rangle=\langle\omega,f\rangle+\int_\Gamma f(\gamma)\operatorname{Re}\frac{1}{\psi(\gamma)}\,d\gamma, \tag{20}$$

which is the desired formula (16).

The support of the measure ω is contained in Γ_0. In fact, for $f\in C_c(\Gamma)$ such that $\operatorname{supp}(f)\cap\Gamma_0=\emptyset$ there exists $n_0\in\mathbb{N}$ such that $\pi^{-1}(U_n)\cap\operatorname{supp}(f)=\emptyset$ for $n\geqq n_0$. It follows that

$$\int_{\pi^{-1}(U_n)} f(\gamma)\operatorname{Re}\frac{1}{\lambda+\psi(\gamma)}\,d\gamma=0 \quad \text{for } \lambda>0 \text{ and } n\geqq n_0,$$

or $\langle\omega_n,f\rangle=0$ for $n\geqq n_0$, which implies that $\langle\omega,f\rangle=0$.

The periodicity group for ψ is Γ_0 (cf. Proposition 8.27), and it follows that

$$\Gamma_0\subseteq\operatorname{per}\left(\operatorname{Re}\frac{1}{\lambda+\psi}\,\omega_\Gamma\right) \quad\text{and}\quad \Gamma_0\subseteq\operatorname{per}\left(\operatorname{Re}\frac{1}{\psi}\,\omega_\Gamma\right),$$

hence by (17)

$$\Gamma_0\subseteq\operatorname{per}(\mathscr{F}_G\nu).$$

This shows by (20) that $\Gamma_0\subseteq\operatorname{per}(\omega)$, and ω is consequently a Haar measure on Γ_0 or the zero-measure.

It remains to prove that $\omega=0$ if and only if κ tends to zero at infinity.

For $f\in C_c(G)$ we find by (20) that

$$\mathscr{F}_G(\nu*f*\tilde{f})=|\mathscr{F}_G f|^2\mathscr{F}_G\nu=|\mathscr{F}_G f|^2\omega+\left[|\mathscr{F}_G f|^2\operatorname{Re}\frac{1}{\psi}\right]\omega_\Gamma,$$

and $|\mathscr{F}_G f|^2\omega+\left[|\mathscr{F}_G f|^2\operatorname{Re}\frac{1}{\psi}\right]\omega_\Gamma$ is thus the positive bounded measure on Γ, whose (co)-Fourier transform is the continuous positive definite function $\nu*f*\tilde{f}$ on G. In particular $\left(|\mathscr{F}_G f|^2\operatorname{Re}\frac{1}{\psi}\right)\omega_\Gamma$ is a bounded measure on Γ, and the function $|\mathscr{F}_G f|^2\operatorname{Re}\frac{1}{\psi}$ is thus integrable for all $f\in C_c(G)$. Furthermore we have that

$$|\mathscr{F}_G f|^2\omega=\mathscr{F}_G((\mathscr{F}_\Gamma\omega)*f*\tilde{f}),$$

and $\mathscr{F}_\Gamma\omega$ is a Haar measure on $G_0=\Gamma_0^\perp$ or the zero measure cf. 6.19. It follows that for $x\in G$ and $f\in C_c(G)$

$$\nu*f*\tilde{f}(x)=(\mathscr{F}_\Gamma\omega)*f*\tilde{f}(x)+\int_\Gamma(x,\gamma)|\mathscr{F}_G f(\gamma)|^2\operatorname{Re}\frac{1}{\psi(\gamma)}\,d\gamma. \tag{21}$$

It is clear that κ tends to zero at infinity if and only if $\nu=\frac{1}{2}(\kappa+\check{\kappa})$ tends to zero at infinity.

Since every function $g\in C_c^+(G)$, can be dominated by a function of the form $f*\tilde{f}$, where $f\in C_c^+(G)$, a positive measure τ on G tends to zero at infinity if and only if $\tau*f*\tilde{f}\in C_0(G)$ for all $f\in C_c^+(G)$.

The second member on the right-hand side of (21) tends to zero at infinity, because $|\mathscr{F}_G f|^2\operatorname{Re}\frac{1}{\psi}$ is integrable on Γ, and it follows that ν tends to zero at infinity if $\omega=0$. If ν tends to zero at infinity, then $\mathscr{F}_\Gamma\omega$ tends to zero at infinity by (21) and the above remark. However, $\mathscr{F}_\Gamma\omega$ is either the zero-measure or a Haar measure on the closed subgroup G_0. Since $(\mu_t)_{t>0}$ is transient and $\psi(0)=0$, G_0 is non-compact (cf. 13.4), and a Haar measure on G_0 does not tend to zero at infinity on G, cf. 1.14 and 1.16. It follows that $\mathscr{F}_\Gamma\omega=0$ and hence $\omega=0$. □

13.34. Corollary. *Let $(\mu_t)_{t>0}$ be an integrable convolution semigroup on G with associated continuous negative definite function ψ on Γ and potential kernel κ. The Fourier transform of the measure $\frac{1}{2}(\kappa+\check{\kappa})$ is the positive definite measure given by*

$$\mathscr{F}_G\left(\frac{1}{2}(\kappa+\check{\kappa})\right)=\operatorname{Re}\frac{1}{\psi}\,\omega_\Gamma.$$

Proof. The result follows from Theorem 13.33 since κ tends to zero at infinity, cf. Proposition 13.21. □

13.35. Corollary. *Let $(\mu_t)_{t>0}$ be a transient and symmetric convolution semigroup on G with associated continuous negative definite function ψ on Γ. The potential kernel κ for $(\mu_t)_{t>0}$ is a symmetric and positive definite measure on G whose Fourier transform is given by*

$$\mathscr{F}_G\kappa=\frac{1}{\psi}\omega_\Gamma.$$

Proof. The result follows from Corollary 13.34 because a transient end symmetric convolution semigroup is integrable, cf. 13.27. □

The measure ω in the representation (16) of the Fourier transform of $\nu=\frac{1}{2}(\kappa+\check{\kappa})$ can be characterized in terms of ν in the following way.

13.36. Theorem (Berg and Forst [2]). *Let $(\mu_t)_{t>0}$ be a transient convolution semigroup on G with potential kernel κ and associated continuous negative definite function ψ on Γ such that $\psi(0)=0$. Let G_0 be the smallest closed subgroup of G containing $\bigcup_{t>0}\operatorname{supp}(\mu_t)$ and ω_0 a fixed Haar measure on G_0. Then the set*

$$I=\{\alpha\geqq 0\mid \nu-\alpha\omega_0\in M_p(G)\},$$

where $\nu=\frac{1}{2}(\kappa+\check{\kappa})$, is a compact interval. Putting $\alpha_0=\sup I$, we have $I=[0,\alpha_0]$, and the measure $\mathscr{F}_G\nu$ is given by

$$\mathscr{F}_G\nu=\alpha_0\mathscr{F}_G\omega_0+\operatorname{Re}\frac{1}{\psi}\omega_\Gamma. \tag{22}$$

The measure $\nu-\alpha_0\omega_0$ tends to zero at infinity on G.

For the *proof* we will need the following result:

13.37. Lemma. *Let φ be a real continuous positive definite function on a non-compact LCA-group G. Then*

$$\limsup_{x\to\infty}\varphi(x)\geqq 0,$$

i.e. for every $\varepsilon>0$ and every compact set $K\subseteq G$ there exists a point $x\notin K$ such that $\varphi(x)>-\varepsilon$.

Proof. Suppose that there exist an $\varepsilon>0$ and a compact set $K\subseteq G$ such that

$$\varphi(x)\leqq-\varepsilon \quad \text{for } x\notin K.$$

Putting $a=\int_K|\varphi(x)|dx$, we may choose a compact set $L\subseteq G$ such that

$$K\subseteq L \quad\text{and}\quad \int_{L\setminus K}dx>\frac{2a}{\varepsilon}.$$

For a function $f\in C_c^+(\Gamma)$ with $\langle\omega_\Gamma, f\rangle=1$ and support contained in a sufficiently small neighbourhood of the neutral element in Γ, we have by Theorem 3.13 that

$$0\leqq|\mathscr{F}_\Gamma f|^2\leqq 1 \quad\text{and}\quad |\mathscr{F}_\Gamma f|^2\geqq\frac{1}{2} \text{ on } L,$$

and denoting by μ the measure associated with $\varphi\in CP(G)$ we find

$$\begin{aligned}
0&\leqq\int_\Gamma f*\tilde{f}(\gamma)\,d\mu(\gamma)=\int_G \varphi(x)|\mathscr{F}_\Gamma f(x)|^2dx\\
&=\int_K \varphi(x)|\mathscr{F}_\Gamma f(x)|^2dx+\int_{G\setminus K}\varphi(x)|\mathscr{F}_\Gamma f(x)|^2dx\\
&\leqq a-\varepsilon\int_{G\setminus K}|\mathscr{F}_\Gamma f(x)|^2dx\\
&\leqq a-\varepsilon\int_{L\setminus K}|\mathscr{F}_\Gamma f(x)|^2dx\\
&<a-\frac{1}{2}\varepsilon\cdot\frac{2a}{\varepsilon}=0,
\end{aligned}$$

which is a contradiction. □

Proof of Theorem 13.36. For $f\in C_c^+(G)$ such that $\langle\omega_0, f*\tilde{f}\rangle=1$ and $\alpha\in I$ we have

$$0\leqq\langle\nu-\alpha\omega_0, f*\tilde{f}\rangle=\langle\nu, f*\tilde{f}\rangle-\alpha,$$

and it follows that I is bounded. With $\alpha_0=\sup I$ it is easy to see that $I=[0,\alpha_0]$.

From Theorem 13.33 we get that

$$\mathscr{F}_G\nu=\beta\mathscr{F}_G\omega_0+\operatorname{Re}\frac{1}{\psi}\omega_\Gamma \tag{23}$$

for some $\beta\geqq 0$, because $\mathscr{F}_G\omega_0$ is a Haar measure on the closed subgroup $\Gamma_0=\{\gamma\in\Gamma\mid\psi(\gamma)=0\}$ of Γ, cf. 6.19 and 8.27.

It follows by (23) that

$$\langle\nu-\beta\omega_0, f*\tilde{f}\rangle=\int|\bar{\mathscr{F}}_G f(\gamma)|^2\operatorname{Re}\frac{1}{\psi(\gamma)}\,d\gamma\geqq 0 \quad\text{for all } f\in C_c(G),$$

and this shows that $\nu-\beta\omega_0$ is a positive definite measure, hence $\beta\leqq\alpha_0$.

Writing $\sigma=\nu-\alpha_0\omega_0$, we get by (23) that for all $f\in C_c(G)$ and $x\in G$

$$\sigma*f*\tilde{f}(x)=(\beta-\alpha_0)\omega_0*f*\tilde{f}(x)+\int(x,\gamma)|\bar{\mathscr{F}}_G f(\gamma)|^2\operatorname{Re}\frac{1}{\psi(\gamma)}\,d\gamma. \tag{24}$$

We now apply (24) to a function $f\in C_c^+(G)$ for which $\langle\omega_0, f*\tilde{f}\rangle=1$. Let $\varepsilon>0$ be given. Since the second member on the right-hand side of (24) tends to zero at infinity, there exists a compact set $K \subseteq G$ such that

$$\left| \int (x, \gamma) |\mathscr{F}_G f(\gamma)|^2 \operatorname{Re} \frac{1}{\psi(\gamma)} d\gamma \right| < \frac{\varepsilon}{2} \quad \text{for } x \in G \setminus K.$$

By Lemma 13.37 applied to the restriction of $\sigma * f * \tilde{f}$ to the non-compact group G_0, there exists a point $x_0 \in G_0 \setminus K$ such that

$$\sigma * f * \tilde{f}(x_0) \geqq -\frac{\varepsilon}{2},$$

and since

$$\omega_0 * f * \tilde{f}(x_0) = \langle \omega_0, f * \tilde{f} \rangle = 1,$$

it follows that

$$\beta - \alpha_0 = \sigma * f * \tilde{f}(x_0) - \int (x_0, \gamma) |\mathscr{F}_G f(\gamma)|^2 \operatorname{Re} \frac{1}{\psi(\gamma)} d\gamma \geqq -\varepsilon,$$

hence $\beta \geqq \alpha_0$. This shows that $\beta = \alpha_0$ and (22) follows.

By (23) we get for all $f \in C_c(G)$ and $x \in G$ that

$$(\nu - \alpha_0 \omega_0) * f * \tilde{f}(x) = \int (x, \gamma) |\mathscr{F}_G f(\gamma)|^2 \operatorname{Re} \frac{1}{\psi(\gamma)} d\gamma,$$

and it follows that

$$\lim_{x \to \infty} (\nu - \alpha_0 \omega_0) * f * \tilde{f}(x) = 0 \quad \text{for all } f \in C_c(G),$$

hence by polarization

$$\lim_{x \to \infty} (\nu - \alpha_0 \omega_0) * f * g(x) = 0 \quad \text{for all } f, g \in C_c(G). \tag{25}$$

Let $f \in C_c(G)$ and $\varepsilon > 0$ be given. The measure $\nu - \alpha_0 \omega_0$ being shift-bounded, cf. 13.10, there exists by Proposition 1.19 a neighbourhood V of the neutral element in G such that for $x, y \in G$ with $x - y \in V$

$$|(\nu - \alpha_0 \omega_0) * f(x) - (\nu - \alpha_0 \omega_0) * f(y)| \leqq \varepsilon.$$

Let $g \in C_c^+(G)$ be such that $\operatorname{supp}(g) \subseteq V$ and $\langle \omega_G, g \rangle = 1$. We then have

$$|(\nu - \alpha_0 \omega_0) * f(x) - (\nu - \alpha_0 \omega_0) * f * g(x)| \leqq \varepsilon \quad \text{for } x \in G,$$

and it follows then easily that

$$\lim_{x \to \infty} (\nu - \alpha_0 \omega_0) * f(x) = 0. \quad \square$$

13.38. Remarks. 1) The constant α_0 in (22) is the uniquely determined number $\alpha \geqq 0$ such that the measure $\nu - \alpha\omega_0$ tends to zero at infinity.

2) In the case $G_0 = G$ we have

$$\lim_{x \to \infty} \nu * f(x) = \alpha_0 \langle \omega_G, f \rangle \qquad \text{for all } f \in C_c(G).$$

13.39. Example. The Poisson semigroup $(\mu_t)_{t>0}$ on $\mathbb{R}$ with jump $s=1$ (cf. 10.2) is transient and the potential kernel for $(\mu_t)_{t>0}$ is the measure

$$\kappa = \sum_{n=0}^{\infty} \varepsilon_n,$$

which is equal to the elementary kernel determined by ε_1 (cf. 13.6). Clearly κ does not tend to zero at infinity, and $(\mu_t)_{t>0}$ is thus not integrable. The measure $\nu = \frac{1}{2}(\kappa + \check{\kappa})$ is

$$\nu = \frac{1}{2} \sum_{n \in \mathbb{Z}} \varepsilon_n + \frac{1}{2} \varepsilon_0.$$

The smallest closed subgroup of $\mathbb{R}$ containing $\bigcup_{t>0} \operatorname{supp}(\mu_t)$ is equal to $\mathbb{Z}$. The associated continuous negative definite function ψ on $\mathbb{R}$ is

$$\psi(y) = 1 - e^{-iy} \qquad \text{for } y \in \mathbb{R},$$

and the periodicity group for ψ is clearly $2\pi\mathbb{Z}$, which is the orthogonal complement of $\mathbb{Z}$. For $y \notin 2\pi\mathbb{Z}$ we find

$$\operatorname{Re} \frac{1}{\psi(y)} = \frac{1 - \cos y}{(1 - \cos y)^2 + \sin^2 y} = \frac{1}{2}.$$

As Haar measure $\omega_{\mathbb{Z}}$ on $\mathbb{Z}$ we use $\sum_{n \in \mathbb{Z}} \varepsilon_n$, and by 13.38 we have $\alpha_0 = \frac{1}{2}$ because the measure $\nu - \alpha\,\omega_{\mathbb{Z}} = \frac{1}{2}\varepsilon_0 + (\frac{1}{2} - \alpha) \sum_{n \in \mathbb{Z}} \varepsilon_n$ tends to zero at infinity precisely for $\alpha = \frac{1}{2}$.

By Example 6.22 we have

$$\mathscr{F}_{\mathbb{R}}(\omega_{\mathbb{Z}}) = \sum_{n \in \mathbb{Z}} \varepsilon_{2\pi n},$$

and it follows that

$$\mathscr{F}_{\mathbb{R}}(\nu) = \frac{1}{2} \sum_{n \in \mathbb{Z}} \varepsilon_{2\pi n} + \frac{1}{2}\,\frac{1}{2\pi}\,\omega_{\mathbb{R}},$$

where $\omega_{\mathbb{R}}$ is Lebesgue measure on $\mathbb{R}$, cf. 5.1.

13.40. Exercise. Find the representation (16), in the case of the translation semigroup $(\varepsilon_t)_{t>0}$ on $\mathbb{R}$ and in the case of the translation semigroup $(\varepsilon_t \otimes \varepsilon_0)_{t>0}$ on $\mathbb{R}^2$.

13.41. Exercise. The convolution semigroup $(\mu_t)_{t>0}$ on $\mathbb{R}$ determined by the measure $\mu = \frac{1}{2}(\varepsilon_1 + \varepsilon_2)$ is transient, but not integrable. The potential kernel κ is given by

$$\kappa = \sum_{n=0}^{\infty} a_n \varepsilon_n,$$

where $\alpha_0 = 1$, $\alpha_1 = \frac{1}{2}$ and $a_n = \frac{1}{2}(a_{n-1} + a_{n-2})$, $n \geqq 2$. Find the representation (16) and prove that the function $\operatorname{Re} \dfrac{1}{\psi}$ "extends to" a continuous positive function, which is not positive definite.

13.42. Exercise. Let $(\mu_t)_{t>0}$ be a convolution semigroup on G with resolvent $(\rho_\lambda)_{\lambda>0}$ and let ψ be the associated continuous negative definite function on Γ. Let finally V be an open, relatively compact neighbourhood of the neutral element in Γ.

Prove that the following conditions are equivalent:

(i) $(\mu_t)_{t>0}$ is transient.

(ii) $\lim\limits_{\lambda \to 0} (\rho_\lambda + \check{\rho}_\lambda)$ exists vaguely.

(iii) $\lim\limits_{\lambda \to 0} \operatorname{Re}\left(\dfrac{1}{\psi + \lambda}\right) \omega_\Gamma$ exists vaguely.

(iv) $\limsup\limits_{\lambda \to 0} \int\limits_V \operatorname{Re} \dfrac{1}{\psi(\gamma) + \lambda} \, d\gamma < \infty$.

The equivalence between (i) and (iv) is called the *Chung-Fuchs criterion*. Using the results from 13.14 and 13.17 the above conditions are equivalent with

(v) $\int\limits_V \operatorname{Re} \dfrac{1}{\psi(\gamma)} \, d\gamma < \infty$.

The implication (iv) ⇒ (v) is simple to establish, but there seems to be no direct analytic proof of the implication (v) ⇒ (iv).

13.43. The notion of transience has a probabilistic interpretation in terms of random walks (which lead to elementary kernels) and infinitely divisible stochastic processes on G. As remarked in 13.17 the two cases are essentially equivalent.

The book of Spitzer [1] contains many criteria for transience and integrability of random walks on the group $G = \mathbb{Z}^n$. See also Feller [1]. The work of Port and Stone [2] is a detailed study of recurrent and transient infinitely divisible stochastic processes and in particular complete results about the behaviour at infinity of the potential kernel κ in the transient but non-integrable case (the type II case in the terminology of Port and Stone) are obtained.

§ 14. Transient Convolution Semigroups on the Half-Axis and Integrals of Convolution Semigroups

A convolution semigroup on $\mathbb{R}$ supported by $[0, \infty[$, which is not the degenerate semigroup $(\varepsilon_0)_{t>0}$, is transient, and we shall study the set of potential kernels for such semigroups by means of the set $\mathscr{P}$ of their Laplace transforms. The set $\mathscr{P}$ is not a convex cone, but it turns out that $\mathscr{P}$ contains two interesting convex cones $\mathscr{H}$ and $\mathscr{S}$, and it is possible to characterize the corresponding convex cones of potential kernels.

To a potential kernel κ for a transient convolution semigroup $(\mu_t)_{t>0}$ on the LCA-group G and a potential kernel τ for a transient convolution semigroup $(\eta_t)_{t>0}$ on $\mathbb{R}$ supported by $[0, \infty[$ is associated a potential kernel κ_τ on G. It is the potential kernel for the convolution semigroup $(\mu_t^f)_{t>0}$ subordinated to $(\mu_t)_{t>0}$ by means of $(\eta_t)_{t>0}$, cf. 9.20. The fractional power of a potential kernel κ and the sum of resolvent measures corresponding to a potential kernel κ are examples of potential kernels of the form κ_τ.

14.1. Proposition. *Let $(\eta_t)_{t>0}$ be a convolution semigroup on $\mathbb{R}$ supported by $[0, \infty[$, and let f be the corresponding Bernstein function.*

Either f is identically zero and then $\eta_t = \varepsilon_0$ for all $t>0$, or $f(x)>0$ for all $x>0$ and then $(\eta_t)_{t>0}$ is transient. In the transient case, the Laplace transform of the potential kernel τ for $(\eta_t)_{t>0}$ is given by

$$\mathscr{L}\tau = \frac{1}{f}.$$

Proof. A Bernstein function f is increasing and concave, and therefore f must be identically zero if there exists a point $x>0$ such that $f(x)=0$, and in this case we have $\eta_t = \varepsilon_0$ for all $t>0$.

Suppose now that $f(x)>0$ for all $x>0$. Then for $x>0$

$$\int_0^\infty \mathscr{L}\eta_t(x)\,dt = \int_0^\infty \exp(-t f(x))\,dt = \frac{1}{f(x)}, \tag{1}$$

and for $\varphi \in C_c^+(\mathbb{R})$ there exists a constant $c>0$ such that

$$\varphi(s) \leqq c e^{-s} \quad \text{for } s \geqq 0.$$

We therefore get

$$\int_0^\infty \langle \eta_t, \varphi \rangle\,dt \leqq c \int_0^\infty \mathscr{L}\eta_t(1)\,dt = \frac{c}{f(1)} < \infty,$$

which shows that $(\eta_t)_{t>0}$ is transient. It follows by (1) that the Laplace transform of the potential kernel τ for $(\eta_t)_{t>0}$ is given by

$$\mathscr{L}\tau = \frac{1}{f}. \quad \square$$

14.2. By $\mathscr{P}$ we denote the set of functions g of the form $g=1/f$, where f is a non-zero Bernstein function.

It follows by Proposition 14.1 (or by Exercise 9.9) that $\mathscr{P}$ can be characterized as the set of non-zero completely monotone functions g for which $1/g$ is a Bernstein function.

By Proposition 14.1 it follows furthermore that the Laplace transformation establishes a one-to-one correspondence between the set of potential kernels for transient convolution semigroups on $\mathbb{R}$ supported by $[0, \infty[$ and the set $\mathscr{P}$.

14.3. Exercise. 1) The set $\mathscr{P}$ is a cone but not a convex cone.
2) For $f \in \mathscr{P}$ we have $f+1 \in \mathscr{P}$.
3) For $f, g \in \mathscr{P}$ we have $fg(f+g)^{-1} \in \mathscr{P}$, in particular $f(f+1)^{-1} \in \mathscr{P}$.

14.4. Definition. A sequence $(a_n)_{n \geqq 0}$ of non-negative numbers is called *logarithmic convex*, if

$$a_n^2 \leqq a_{n-1} a_{n+1} \qquad \text{for all integers } n \geqq 1.$$

The set of logarithmic convex sequences is a convex cone. In fact, if $(a_n)_{n \geqq 0}$ and $(b_n)_{n \geqq 0}$ are logarithmic convex sequences, then so is $(a_n+b_n)_{n \geqq 0}$, because

$$\begin{aligned}(a_n+b_n)^2 &= a_n^2+b_n^2+2a_n b_n \\ &\leqq a_{n-1}a_{n+1}+b_{n-1}b_{n+1}+2(a_{n-1}a_{n+1}b_{n-1}b_{n+1})^{\frac{1}{2}} \\ &\leqq a_{n-1}a_{n+1}+b_{n-1}b_{n+1}+a_{n-1}b_{n+1}+a_{n+1}b_{n-1} \\ &= (a_{n-1}+b_{n-1})(a_{n+1}+b_{n+1}).\end{aligned}$$

14.5. By $\mathscr{H}$ we denote the set of all completely monotone functions f, such that for all $x>0$ the sequence

$$\left(\frac{(-1)^n}{n!} D^n f(x)\right)_{n \geqq 0}$$

is logarithmic convex.

Note that for a non-constant completely monotone function f we have

$$\frac{(-1)^n}{n!} D^n f(x) > 0 \qquad \text{for all } x>0 \text{ and } n \geqq 0.$$

The following result about $\mathscr{H}$ is due to Hirsch [3].

14.6. Proposition. *The set $\mathscr{H}$ is a convex cone and a closed subset of $\mathbb{R}^{]0,\infty[}$ in the topology of pointwise convergence.*
The set $\mathscr{H} \setminus \{0\}$ is contained in $\mathscr{P}$.

Proof. It is clear by 14.4 that $\mathscr{H}$ is a convex cone, and the closedness of $\mathscr{H}$ follows immediately by Proposition 9.5 and Remark 9.7.

Let $f \in \mathscr{H} \setminus \{0\}$. If f is (a non-zero) constant, then $f \in \mathscr{P}$, and for the rest of the proof we assume that f is non-constant. For fixed $x > 0$, we consider the numbers

$$a_0 = \frac{1}{f(x)} \quad \text{and} \quad a_n = \frac{(-1)^{n-1}}{n!} D^n \left(\frac{1}{f}\right)(x) \quad \text{for } n \geqq 1,$$

and we shall prove that $a_n \geqq 0$ for $n \geqq 0$. This is clear for $n = 0, 1$, and the proof will proceed by induction. Putting

$$b_n = \frac{(-1)^n}{n!} D^n f(x) \quad \text{for } n \geqq 0,$$

we have, since f is non-constant, that

$$b_n > 0 \quad \text{for } n \geqq 0,$$

and by hypothesis

$$b_n^2 \leqq b_{n-1} b_{n+1} \quad \text{for } n \geqq 1,$$

so it follows that

$$b_n b_{n+1-p} \leqq b_{n-p} b_{n+1} \quad \text{for } p = 1, \ldots, n. \tag{2}$$

Deriving n times the identity $1 = f \cdot \frac{1}{f}$ we find

$$a_0 b_0 = 1 \quad \text{and} \quad a_0 b_n = \sum_{p=1}^{n} a_p b_{n-p} \quad \text{for } n \geqq 1,$$

which gives

$$\sum_{p=1}^{n} a_p \frac{b_{n-p}}{b_n} = \sum_{p=1}^{n+1} a_p \frac{b_{n+1-p}}{b_{n+1}},$$

hence

$$a_{n+1} = \frac{b_{n+1}}{b_0} \sum_{p=1}^{n} a_p \left(\frac{b_{n-p}}{b_n} - \frac{b_{n+1-p}}{b_{n+1}} \right). \tag{3}$$

Suppose that $a_i \geqq 0$ for $i = 0, 1, \ldots, n$. Then by (2) and (3) it follows that $a_{n+1} \geqq 0$, which shows the induction step. □

14.7. Definition. A function $k\colon]0, \infty[\to \mathbb{R}$ is called *logarithmic convex* if either k is identically zero, or if $k(x) > 0$ for all $x > 0$ and $\log k$ is a convex function.

A logarithmic convex function is continuous and convex. The set of logarithmic convex functions is a convex cone and a closed subset of $\mathbb{R}^{]0,\infty[}$ in the topology of pointwise convergence, cf. e.g. Artin [1].

A function $k\colon]0, \infty[\to]0, \infty[$ of class C^2 is logarithmic convex if and only if $(Dk)^2 \leqq k D^2 k$. It follows that a completely monotone function f is logarithmic convex, because the inequality $(Df)^2 \leqq f D^2 f$ is an immediate consequence of the Cauchy-Schwarz inequality applied to the representation 9.3 (1) for f.

The following theorem due to Hirsch [4], characterizes the set of positive measures on $[0,\infty[$ for which the Laplace transform belongs to $\mathscr{H}$.

14.8. Theorem. *Let $k\colon]0,\infty[\to\mathbb{R}$ be a decreasing logarithmic convex function such that the function*

$$k_0(x)=\begin{cases} k(x) & \text{for } x>0,\\ 0 & \text{for } x\leqq 0,\end{cases}$$

is locally integrable on $\mathbb{R}$ and let $a\geqq 0$.

The Laplace transform of the measure

$$\tau=a\varepsilon_0+k_0(x)\,dx \tag{4}$$

belongs to $\mathscr{H}$, and every element of $\mathscr{H}$ is the Laplace transform of a measure τ of this form.

Proof. Suppose first that k can be extended to a function of class C^2 on $[0,\infty[$. For every $s>0$ the function g_s, defined by

$$g_s(x)=e^{-sx}k(x) \qquad \text{for } x>0,$$

is also logarithmic convex and of class C^2 and hence

$$(Dg_s)^2\leqq g_s D^2 g_s.$$

Putting

$$f(s)=\mathscr{L}k(s)=\int_0^\infty e^{-sx}k(x)\,dx \qquad \text{for } s>0,$$

we find for $s>0$ and $n\geqq 1$ that

$$\frac{(-1)^n}{n!}D^n f(s)=\int_0^\infty \frac{x^n}{n!}g_s(x)\,dx=-\int_0^\infty \frac{x^{n+1}}{(n+1)!}Dg_s(x)\,dx.$$

By the Cauchy-Schwarz inequality we then find

$$\begin{aligned}\left(\frac{(-1)^n}{n!}D^n f(s)\right)^2 &\leqq \int_0^\infty \frac{x^{n+1}}{(n+1)!}g_s(x)\,dx\int_0^\infty \frac{x^{n+1}}{(n+1)!}D^2 g_s(x)\,dx\\ &=\int_0^\infty \frac{x^{n+1}}{(n+1)!}g_s(x)\,dx\int_0^\infty \frac{x^{n-1}}{(n-1)!}g_s(x)\,dx\\ &=\frac{(-1)^{n+1}}{(n+1)!}D^{n+1}f(s)\frac{(-1)^{n-1}}{(n-1)!}D^{n-1}f(s),\end{aligned}$$

which shows that $f\in\mathscr{H}$.

Every decreasing logarithmic convex function $k\colon]0,\infty[\to]0,\infty[$ is the pointwise limit of an increasing sequence of decreasing, logarithmic convex functions k_n: $]0,\infty[\to]0,\infty[$ of the above type, i.e. for which $\mathscr{L}k_n \in \mathscr{H}$. By the monotone convergence theorem it follows that $\mathscr{L}k_n$ increases pointwise to $\mathscr{L}k$, so by Proposition 14.6 we get that $\mathscr{L}k \in \mathscr{H}$.

Since the positive constant functions belong to the convex cone $\mathscr{H}$ it follows that $\mathscr{L}\tau \in \mathscr{H}$.

Consider now a function $f \in \mathscr{H}$. Since f is completely monotone there exists a positive measure τ on $[0,\infty[$ such that $f = \mathscr{L}\tau$. We shall prove that τ has the form (4), and this is clear, when f is constant, so we will assume that f is non-constant and then we have, cf. 14.5, that

$$\frac{(-1)^n}{n!} D^n f(x) > 0 \qquad \text{for } x>0 \text{ and } n \geqq 0.$$

For each $n \in \mathbb{N}$ we define a function $L_{n,f}$: $]0,\infty[\to\mathbb{R}$ by

$$L_{n,f}(x) = \frac{(-1)^n}{n!} D^n f\left(\frac{n}{x}\right)\left(\frac{n}{x}\right)^{n+1} \qquad \text{for } x>0,$$

and proceed in three steps:

a) $L_{n,f}$ is logarithmic convex.
The function $L_{n,f}$ is strictly positive and of class C^∞, so it suffices to prove that $D^2(\log L_{n,f}) \geqq 0$. We find

$$D^2(\log L_{n,f})(x) = \frac{n+1}{x^2} + \frac{2n}{x^3} \frac{D^{n+1} f\left(\frac{n}{x}\right)}{D^n f\left(\frac{n}{x}\right)}$$

$$+ \left(\frac{n}{x^2}\right)^2 \frac{D^{n+2} f\left(\frac{n}{x}\right)}{D^n f\left(\frac{n}{x}\right)} - \left(\frac{n}{x^2} \frac{D^{n+1} f\left(\frac{n}{x}\right)}{D^n f\left(\frac{n}{x}\right)}\right)^2,$$

and using the inequality

$$D^n f\left(\frac{n}{x}\right) D^{n+2} f\left(\frac{n}{x}\right) \geqq \frac{n+2}{n+1}\left(D^{n+1} f\left(\frac{n}{x}\right)\right)^2,$$

we get

$$D^2(\log L_{n,f})(x) \geqq \frac{n+1}{x^2} + \frac{2n}{x^3} \frac{D^{n+1} f\left(\frac{n}{x}\right)}{D^n f\left(\frac{n}{x}\right)} + \frac{1}{n+1}\left(\frac{n}{x^2} \frac{D^{n+1} f\left(\frac{n}{x}\right)}{D^n f\left(\frac{n}{x}\right)}\right)^2$$

$$= \left(\frac{\sqrt{n+1}}{x} + \frac{D^{n+1} f\left(\frac{n}{x}\right)}{D^n f\left(\frac{n}{x}\right)} \frac{n}{x^2\sqrt{n+1}}\right)^2 \geqq 0.$$

b) $L_{n,f}$ is decreasing.

We have that

$$f(s)=\int_0^\infty e^{-st}d\tau(t),$$

and if we suppose that τ is bounded (or equivalently that $\lim_{s\to 0} f(s)<\infty$), it is easy to see that

$$\lim_{s\to 0} D^n f(s) s^{n+1}=0 \quad \text{for } n\geqq 0,$$

and hence

$$\lim_{x\to\infty} L_{n,f}(x)=0.$$

The function $L_{n,f}$, being convex and positive, is then necessarily decreasing.

In the general case we consider the function f_h for $h>0$, defined by

$$f_h(s)=f(s+h).$$

The function f_h belongs to $\mathscr{H}$ and the representing measure $e^{-ht}d\tau(t)$ is bounded, so that L_{n,f_h} is decreasing. Since

$$\lim_{h\to 0} L_{n,f_h}(x)=L_{n,f}(x) \quad \text{for } x>0,$$

it follows that $L_{n,f}$ is decreasing.

c) Putting $a=\lim_{s\to\infty} f(s)$ we have

$$\tau=a\varepsilon_0+\tau_0,$$

where τ_0 is the restriction of τ to $]0,\infty[$. However

$$\lim_{n\to\infty} L_{n,f}(x)dx=\tau_0$$

vaguely as measures on $]0,\infty[$, (this can be seen e.g. by means of Theorem 5b p. 287 in Widder [1]), and since (by a) and b)) $D(L_{n,f})\leqq 0$ and $D^2(L_{n,f})\geqq 0$, we get that $D(\tau_0)\leqq 0$ and $D^2(\tau_0)\geqq 0$ as distributions on $]0,\infty[$. It follows, cf. Schwartz [1] p. 54, that there exists a convex decreasing function k on $]0,\infty[$ such that $\tau_0=k(x)dx$, and k is in particular continuous, so by Widder [1], p. 288 we get

$$\lim_{n\to\infty} L_{n,f}(x)=k(x)$$

for all $x>0$. By a) we finally obtain that k is logarithmic convex. □

14.9. Corollary. *Every non-zero measure τ on $[0,\infty[$ of the form* (4) *is potential kernel for a transient convolution semigroup on* $\mathbb{R}$ *supported by* $[0,\infty[$.

Proof. This follows immediately from Theorem 14.8 and Proposition 14.6. □

A very important subset of $\mathscr{H}$ is formed by the functions, which are Stieltjes transforms.

14.10. Definition. A function $f\colon]0,\infty[\to\mathbb{R}$ is called a *Stieltjes transform*, if there exist a constant $a\geqq 0$ and a positive measure μ on $[0,\infty[$ such that

$$f(x)=a+\int_0^\infty \frac{1}{x+t}\,d\mu(t) \quad \text{for all } x>0. \tag{5}$$

The set of Stieltjes transforms is denoted $\mathscr{S}$.

The pair (a,μ) associated with a Stieltjes transform f is uniquely determined. In fact

$$a=\lim_{x\to\infty} f(x),$$

and

$$\begin{aligned}\int_0^\infty \frac{1}{x+t}\,d\mu(t)&=\int_0^\infty\left(\int_0^\infty e^{-(x+t)u}\,du\right)d\mu(t)\\&=\int_0^\infty e^{-xu}\mathscr{L}\mu(u)\,du=\mathscr{L}(\mathscr{L}\mu)(x),\end{aligned} \tag{6}$$

and the unicity of μ follows by the injectivity of the Laplace transformation. Cf. 9.11.

14.11. Proposition. *The set $\mathscr{S}$ of Stieltjes transforms is a convex cone contained in $\mathscr{H}$.*

Proof. It follows immediately from (6) that a Stieltjes transform f is completely monotone, and for $n\geqq 1$ and $x>0$ we find

$$\frac{(-1)^n}{n!}D^n f(x)=\int_0^\infty \frac{1}{(x+t)^{n+1}}\,d\mu(t),$$

and it follows by the Cauchy-Schwarz inequality that the sequence

$$\left(\frac{(-1)^n}{n!}D^n f(x)\right)_{n\geqq 0}$$

is logarithmic convex. This proves that $\mathscr{S}\subseteq\mathscr{H}$, and $\mathscr{S}$ is clearly a convex cone. □

14.12. Remark. The cone $\mathscr{S}$ has the following properties:

(i) $\mathscr{S}$ is a closed subset of $\mathbb{R}^{]0,\infty[}$ in the topology of pointwise convergence.

(ii) For $f\in\mathscr{S}$ we have $\dfrac{f}{f+1}\in\mathscr{S}$.

(iii) For $f\in\mathscr{S}\setminus\{0\}$ the function $x\mapsto\dfrac{1}{f\left(\frac{1}{x}\right)}$ belongs to $\mathscr{S}$.

For the proof of these results see Hirsch [3].

14.13. Proposition. *Let g be a completely monotone function on $]0,\infty[$ such that the function*

$$g_0(x)=\begin{cases} g(x) & \text{for } x>0,\\ 0 & \text{for } x\leqq 0,\end{cases}$$

is locally integrable on $\mathbb{R}$ and let $a\geqq 0$.

The Laplace transform of the measure

$$\tau=a\varepsilon_0+g_0(x)\,dx \tag{7}$$

belongs to $\mathscr{S}$ and every element of $\mathscr{S}$ is the Laplace transform of a measure τ of this form.

Proof. The Laplace transform of the measure τ is

$$\mathscr{L}\tau(s)=a+\int_0^\infty e^{-sx}g(x)\,dx,$$

but g is the Laplace transform of a positive measure μ on $[0,\infty[$, and it follows that $\mathscr{L}\tau\in\mathscr{S}$, cf. (6).

If $f\in\mathscr{S}$ there exist a constant $a\geqq 0$ and a positive measure μ on $[0,\infty[$ such that (5) holds, and putting $g=\mathscr{L}\mu$, it follows by (6) that the function

$$g_0(x)=\begin{cases} g(x) & \text{for } x>0,\\ 0 & \text{for } x\leqq 0,\end{cases}$$

is locally integrable and that the Laplace transform of the measure

$$\tau=a\varepsilon_0+g_0(x)\,dx$$

is equal to f. □

14.14. Corollary. *Let g be a non-zero completely monotone function such that the function*

$$g_0(x)=\begin{cases} g(x) & \text{for } x>0,\\ 0 & \text{for } x\leqq 0,\end{cases}$$

is locally integrable on $\mathbb{R}$. For every $a\geqq 0$ the measure

$$\tau=a\varepsilon_0+g_0(x)\,dx$$

is potential kernel for a transient convolution semigroup on $\mathbb{R}$ supported by $[0,\infty[$.

Proof. The function $\mathscr{L}\tau$ belongs to $\mathscr{S}\setminus\{0\}$ so by 14.6 and 14.11 we have that $\mathscr{L}\tau\in\mathscr{P}$. □

14.15. Example. For $\alpha \in [0, 1]$ the function $f(x) = x^{-\alpha}$ for $x > 0$, belongs to $\mathscr{S}$.

This is clear for $\alpha = 0, 1$, and for $\alpha \in]0, 1[$ it follows from the elementary formula

$$x^{-\alpha} = \frac{\sin(\alpha\pi)}{\pi} \int_0^\infty \frac{t^{-\alpha}}{x+t} dt \quad \text{for } x > 0.$$

14.16. Example. The Bernstein function corresponding to the one-sided stable semigroup $(\sigma_t^\alpha)_{t>0}$ of order $\alpha \in [0, 1]$, (cf. 9.23) is $f(x) = x^\alpha$.

In order to describe the potential kernel for the transient convolution semigroups $(\sigma_t^\alpha)_{t>0}$, we consider for every $\alpha > 0$ the function

$$x \mapsto \begin{cases} \dfrac{1}{\Gamma(\alpha)} x^{\alpha-1} & \text{for } x > 0, \\ 0 & \text{for } x \leqq 0, \end{cases} \tag{8}$$

which is locally integrable on $\mathbb{R}$, hence density for a positive measure τ_α on $\mathbb{R}$ for which $\operatorname{supp}(\tau_\alpha) = [0, \infty[$. We define $\tau_0 = \varepsilon_0$. The Laplace transform of τ_α is given by

$$\mathscr{L}\tau_\alpha(s) = \frac{1}{\Gamma(\alpha)} \int_0^\infty e^{-xs} x^{\alpha-1} dx = s^{-\alpha} \quad \text{for } s, \alpha > 0,$$

and it follows that

$$\tau_\alpha * \tau_\beta = \tau_{\alpha+\beta} \quad \text{for } \alpha, \beta \geqq 0, \tag{9}$$

and furthermore by Proposition 14.1 that τ_α for $\alpha \in [0, 1]$ is the potential kernel for $(\sigma_t^\alpha)_{t>0}$. Note that the function $x \mapsto \dfrac{1}{\Gamma(\alpha)} x^{\alpha-1}$ is completely monotone for $\alpha \in]0, 1]$.

Since $\lim_{x\to\infty} x^{\alpha-1} = 0$ for $\alpha \in]0, 1[$, it follows that the measure τ_α tends to zero at infinity for $\alpha \in]0, 1[$ (and also for $\alpha = 0$). By 13.21 we then have that $(\sigma_t^\alpha)_{t>0}$ is integrable for $\alpha \in [0, 1[$ but not for $\alpha = 1$.

14.17. Example. The Γ-semigroup $(\eta_t)_{t>0}$ (cf. 9.23) is transient and the potential kernel is the measure $g(x)\,dx$, where

$$g(x) = \begin{cases} e^{-x} \displaystyle\int_0^\infty \frac{x^{t-1}}{\Gamma(t)} dt & \text{for } x > 0, \\ 0 & \text{for } x \leqq 0. \end{cases}$$

The function $\log\left(1 + \dfrac{1}{x}\right)$ belongs to $\mathscr{S}$ because

$$\log\left(1 + \frac{1}{x}\right) = \int_0^1 \frac{dt}{x+t} \quad \text{for } x > 0,$$

and the function $\dfrac{1}{\log(1+x)}$ belongs therefore by 14.12(iii) to $\mathscr{S}$. It follows that the function $g|]0,\infty[$ is completely monotone. Since, as is easily seen, $\lim_{x\to\infty} g(x)=1$, the Γ-semigroup is not integrable.

Let $(\mu_t)_{t>0}$ be a convolution semigroup on the LCA-group G and let $(\eta_t)_{t>0}$ be a convolution semigroup on $\mathbb{R}$ supported by $[0,\infty[$ with corresponding Bernstein function f.

We shall now study conditions under which the convolution semigroup $(\mu_t^f)_{t>0}$ on G subordinated to $(\mu_t)_{t>0}$ by means of $(\eta_t)_{t>0}$, cf. 9.20, is transient.

If $\eta_t=\varepsilon_0$ for $t>0$ then $\mu_t^f=\varepsilon_0$ for $t>0$. If $\eta_t\neq\varepsilon_0$ for $t>0$ then $(\eta_t)_{t>0}$ is transient (cf. 14.1), and denoting by τ the potential kernel for $(\eta_t)_{t>0}$ we have the following necessary and sufficient condition for the transience of $(\mu_t^f)_{t>0}$.

14.18. Proposition. *The convolution semigroup $(\mu_t^f)_{t>0}$ subordinated to $(\mu_t)_{t>0}$ by means of $(\eta_t)_{t>0}$ is transient if and only if the vague integral*

$$\int_0^\infty \mu_s \, d\tau(s) \tag{10}$$

exists, and the potential kernel for $(\mu_t^f)_{t>0}$ is in the affirmative case the measure determined by (10).

Proof. For $h\in C_c^+(G)$ we find using 9.21 that

$$\int_0^\infty \langle \mu_t^f, h\rangle \, dt = \int_0^\infty \left(\int_0^\infty \langle \mu_s, h\rangle \, d\eta_t(s)\right) dt = \int_0^\infty \langle \mu_s, h\rangle \, d\tau(s),$$

and the assertion follows. □

14.19. Lemma. *Let $(\mu_t)_{t>0}$ be a transient convolution semigroup on G with potential kernel κ. For every positive shift-bounded measure τ on $\mathbb{R}$ such that* $\operatorname{supp}(\tau)\subseteq[0,\infty[$ *the vague integral*

$$\kappa_\tau = \int_0^\infty \mu_t \, d\tau(t) \tag{11}$$

exists. The measure κ_τ on G is shift-bounded and it tends to zero at infinity if κ tends to zero at infinity.

Proof. Putting $c=\sup_{n\geqq 0}\tau([n,n+1])$, we have by Exercise 1.17 that $c<\infty$. The vague integral

$$\sigma=\int_0^1 \mu_t \, dt,$$

clearly defines a positive non-zero bounded measure on G.

For every $h\in C_c^+(G)$ there exists by Lemma 1.9 a function $g\in C_c^+(G)$ such that

$$h\leqq \sigma * g,$$

and for $n \geqq 0$, $s \in [n, n+1]$ and $x \in G$ we then have

$$\mu_s * h(x) \leqq \mu_s * \sigma * g(x) = \int_0^1 \mu_s * \mu_t * g(x)\, dt = \int_s^{s+1} \mu_t * g(x)\, dt \leqq \int_n^{n+2} \mu_t * g(x)\, dt,$$

and hence

$$\int_n^{n+1} \mu_s * h(x)\, d\tau(s) \leqq \tau([n, n+1]) \max_{s \in [n, n+1]} \mu_s * h(x) \leqq c \int_n^{n+2} \mu_t * g(x)\, dt.$$

It follows that

$$\begin{aligned} \int_0^\infty \mu_s * h(x)\, d\tau(s) &= \sum_{n=0}^\infty \int_n^{n+1} \mu_s * h(x)\, d\tau(s) \\ &\leqq c \sum_{n=0}^\infty \int_n^{n+2} \mu_t * g(x)\, dt \\ &\leqq 2c \int_0^\infty \mu_t * g(x)\, dt \\ &= 2c\kappa * g(x). \end{aligned}$$

In particular we have, putting $x = 0$,

$$\int_0^\infty \langle \mu_s, \check{h} \rangle\, d\tau(s) \leqq 2c \langle \kappa, \check{g} \rangle < \infty,$$

which shows that the vague integral (11) exists, and furthermore that

$$\kappa_\tau * h \leqq 2c\kappa * g.$$

The measure κ being shift-bounded (cf. 13.10), it follows that κ_τ is shift-bounded, and κ_τ tends to zero at infinity if κ tends to zero at infinity. □

14.20. Remark. The notation κ_τ in Lemma 14.19 is justified by the fact that κ is potential kernel for exactly one convolution semigroup, cf. 15.21.

The following result is due to Hirsch [4] and Itô [4].

14.21. Theorem. *Let $(\mu_t)_{t>0}$ be a transient convolution semigroup on G with potential kernel κ and let $(\eta_t)_{t>0}$ be a convolution semigroup on $\mathbb{R}$ supported by $[0, \infty[$ such that $\eta_t \neq \varepsilon_0$ for $t > 0$ and with potential kernel τ.*

Then the convolution semigroup $(\mu_t^f)_{t>0}$ subordinated to $(\mu_t)_{t>0}$ by means of $(\eta_t)_{t>0}$ is transient with potential kernel

$$\kappa_\tau = \int_0^\infty \mu_t\, d\tau(t).$$

If $(\mu_t)_{t>0}$ is integrable then $(\mu_t^f)_{t>0}$ is integrable.

Proof. The measure τ on $\mathbb{R}$ is shift-bounded (cf. 13.10) and $\operatorname{supp}(\tau)\subseteq[0,\infty[$, and it follows by Lemma 14.19 that the vague integral

$$\kappa_\tau=\int_0^\infty \mu_t\,d\tau(t)$$

exists. This implies, by Proposition 14.18, that $(\mu_t^f)_{t>0}$ is transient with κ_τ as potential kernel.

The last statement of the theorem follows immediately by Proposition 13.21 and Lemma 14.19. □

Consider the set M of positive shift-bounded measures τ on $\mathbb{R}$ such that $\operatorname{supp}(\tau)\subseteq[0,\infty[$ and with the property that for every LCA-group G and for every potential kernel κ on G the measure κ_τ is a potential kernel on G. The following characterization of M is due to Itô [3].

14.22. Proposition. *The set M is exactly the set of potential kernels for transient convolution semigroups on $\mathbb{R}$ supported by $[0,\infty[$.*

Proof. It follows by Theorem 14.21 that the potential kernels for transient convolution semigroups on $\mathbb{R}$ supported by $[0,\infty[$ belong to M. In order to see that every $\tau\in M$ is such a potential kernel we consider the potential kernel $\kappa=1_{]0,\infty[}(x)\,dx$ for the translation semigroup $(\varepsilon_t)_{t>0}$ on $\mathbb{R}$ and we get that $\tau=\kappa_\tau$ is a potential kernel on $\mathbb{R}$. □

14.23. Definition. By the *potential kernels subordinated* to a given potential kernel κ on G we understand the potential kernels κ_τ, where τ is a potential kernel for a transient convolution semigroup on $\mathbb{R}$ supported by $[0,\infty[$.

The set of potential kernels subordinated to a given potential kernel κ on G is denoted $\mathscr{P}(\kappa)$. We also introduce the subsets

$$\mathscr{H}(\kappa)=\{\kappa_\tau\in\mathscr{P}(\kappa)\mid \mathscr{L}\tau\in\mathscr{H}\setminus\{0\}\}$$

and

$$\mathscr{S}(\kappa)=\{\kappa_\tau\in\mathscr{P}(\kappa)\mid \mathscr{L}\tau\in\mathscr{S}\setminus\{0\}\}.$$

The set $\mathscr{P}(\kappa)$ is in general not a convex cone, but the subsets $\mathscr{H}(\kappa)$ and $\mathscr{S}(\kappa)$ are convex cones of potential kernels subordinated to κ.

14.24. Exercise. For potential kernels τ' and τ'' on $\mathbb{R}$ supported by $[0,\infty[$ the potential kernel $\tau'_{\tau''}$ on $\mathbb{R}$ is also supported by $[0,\infty[$. In this way there is defined an associative composition in the set of potential kernels on $\mathbb{R}$ supported by $[0,\infty[$. The potential kernel $1_{]0,\infty[}(x)\,dx$ is neutral element for this composition. If $g'=\mathscr{L}\tau'$, $g''=\mathscr{L}\tau''$ we have $\mathscr{L}(\tau'_{\tau''})=g''\circ\dfrac{1}{g'}$.

14.25. Let $(\mu_t)_{t>0}$ be a transient convolution semigroup on G with potential kernel κ and associated negative definite function ψ. For $\alpha\in[0,1]$ the convolution semigroup with associated negative definite function ψ^α is (cf. 9.23)

$$\left(\int_0^\infty \mu_s \, d\sigma_t^\alpha(s)\right)_{t>0}$$

By Theorem 14.21 this semigroup is transient and the potential kernel, which is denoted κ^α, is called the *fractional power of* κ *of order* α. We have $\kappa^0 = \varepsilon_0$, $\kappa^1 = \kappa$ and κ^α is for $\alpha \in]0, 1[$ the vague integral, cf. 14.16,

$$\kappa^\alpha = \frac{1}{\Gamma(\alpha)} \int_0^\infty \mu_t t^{\alpha-1} \, dt = \int_0^\infty \mu_t \, d\tau_\alpha(t). \tag{12}$$

It follows that $\kappa^\alpha \in \mathscr{S}(\kappa)$ because $\mathscr{L}\tau_\alpha \in \mathscr{S} \setminus \{0\}$, cf. 14.15 and 14.16.

14.26. Proposition. *Let* κ *be the potential kernel for a transient convolution semigroup* $(\mu_t)_{t>0}$ *on* G. *The mapping* $\alpha \mapsto \kappa^\alpha$ *of* $[0, 1]$ *into* $M^+(G)$ *is vaguely continuous and for* $\alpha, \beta \in [0, 1]$ *such that* $\alpha + \beta \in [0, 1]$ *the convolution of* κ^α *and* κ^β *exists and verifies*

$$\kappa^\alpha * \kappa^\beta = \kappa^{\alpha+\beta}. \tag{13}$$

Proof. Let $f \in C_c^+(G)$. By the monotone convergence theorem it follows that each of the mappings

$$\alpha \mapsto \int_0^1 \langle \mu_t, f \rangle t^{\alpha-1} \, dt$$

and

$$\alpha \mapsto \int_1^\infty \langle \mu_t, f \rangle t^{\alpha-1} \, dt$$

is continuous for $\alpha \in]0, 1]$, and therefore $\alpha \mapsto \kappa^\alpha$ is vaguely continuous on $]0, 1]$.

To see that $\lim_{\alpha \to 0} \langle \kappa^\alpha, f \rangle = \langle \varepsilon_0, f \rangle$ we first remark that

$$0 \leq \lim_{\alpha \to 0} \frac{1}{\Gamma(\alpha)} \int_1^\infty \langle \mu_t, f \rangle t^{\alpha-1} \, dt \leq \lim_{\alpha \to 0} \frac{1}{\Gamma(\alpha)} \int_1^\infty \langle \mu_t, f \rangle \, dt = 0,$$

because $\lim_{\alpha \to 0} \dfrac{1}{\Gamma(\alpha)} = 0$.

For a continuously differentiable function $\varphi \colon [0, 1] \to \mathbb{R}$ we find by partial integration

$$\frac{1}{\Gamma(\alpha)} \int_0^1 \varphi(t) t^{\alpha-1} dt = \frac{1}{\Gamma(\alpha+1)} \left(\varphi(1) - \int_0^1 t^\alpha D\varphi(t) dt\right),$$

and we therefore have

$$\lim_{\alpha \to 0} \frac{1}{\Gamma(\alpha)} \int_0^1 \varphi(t) t^{\alpha-1} \, dt = \varphi(1) - \int_0^1 D\varphi(t) \, dt = \varphi(0).$$

By approximating the function $t\mapsto\langle\mu_t,f\rangle$ uniformly on the interval $[0,1]$ by continuously differentiable functions, it is now easy to see that

$$\lim_{\alpha\to 0}\frac{1}{\Gamma(\alpha)}\int_0^1\langle\mu_t,f\rangle t^{\alpha-1}\,dt=\langle\varepsilon_0,f\rangle,$$

hence

$$\lim_{\alpha\to 0}\langle\kappa^\alpha,f\rangle=\langle\varepsilon_0,f\rangle.$$

For $f\in C_c^+(G)$ and $\alpha,\beta\in]0,1]$ such that $\alpha+\beta\leqq 1$ we find, with the notation from 14.16,

$$\begin{aligned}\int_G\int_G f(x+y)\,d\kappa^\alpha(x)\,d\kappa^\beta(y)&=\int_0^\infty\int_G\int_0^\infty\int_G f(x+y)\,d\mu_t(x)\,d\tau_\alpha(t)\,d\mu_s(y)\,d\tau_\beta(s)\\&=\int_0^\infty\int_0^\infty\int_G\int_G f(x+y)\,d\mu_t(x)\,d\mu_s(y)\,d\tau_\alpha(t)\,d\tau_\beta(s)\\&=\int_0^\infty\int_0^\infty\langle\mu_{t+s},f\rangle\,d\tau_\alpha(t)\,d\tau_\beta(s)=\int_0^\infty\langle\mu_r,f\rangle\,d(\tau_\alpha*\tau_\beta)(r)\\&=\int_0^\infty\langle\mu_r,f\rangle\,d\tau_{\alpha+\beta}(r)=\langle\kappa^{\alpha+\beta},f\rangle,\end{aligned}$$

which shows that the convolution $\kappa^\alpha*\kappa^\beta$ exists and that (13) holds. □

A special case of the following result was first proved by Itô [1].

14.27. Proposition. *Let κ be the potential kernel for a transient convolution semigroup $(\mu_t)_{t>0}$ on G. For every non-zero positive measure μ on $[0,1]$ the measure*

$$\int_0^1\kappa^\alpha\,d\mu(\alpha)$$

is a potential kernel on G belonging to $\mathscr{S}(\kappa)$.

Proof. Writing $\mu=a\varepsilon_0+\mu_0$, where μ_0 is the restriction of μ to $]0,1]$ and $a=\mu(\{0\})\geqq 0$, and defining

$$g(t)=\int_0^1\frac{t^{\alpha-1}}{\Gamma(\alpha)}\,d\mu_0(\alpha)\qquad\text{for } t>0,$$

we have

$$\int_0^1\kappa^\alpha\,d\mu(\alpha)=a\varepsilon_0+\int_0^\infty\mu_t\,g(t)\,dt.$$

The function g is completely monotone since $t\mapsto t^{\alpha-1}$ is so for $\alpha\in]0,1]$, and the function

$$g_0(t)=\begin{cases}g(t)&\text{for } t>0,\\0&\text{for } t\leqq 0,\end{cases}$$

is locally integrable. By 14.13 and 14.14 the measure

$$\tau = a\varepsilon_0 + g_0(t)\,dt$$

is a potential kernel on $\mathbb{R}$ supported by $[0, \infty[$ such that $\mathscr{L}\tau \in \mathscr{S}\setminus\{0\}$, and since we have

$$\int_0^1 \kappa^\alpha \, d\mu(\alpha) = \int_0^\infty \mu_t \, d\tau(t)$$

the assertion follows by Theorem 14.21. □

14.28. Proposition. *Let κ be the potential kernel for a transient convolution semigroup $(\mu_t)_{t>0}$ on G with resolvent $(\rho_\lambda)_{\lambda>0}$ and put $\rho_0 = \kappa$.*

For every non-zero positive measure μ on $[0, \infty[$ such that the vague integral

$$\int_0^\infty \rho_\lambda \, d\mu(\lambda) \tag{14}$$

exists, the measure (14) *is a potential kernel belonging to $\mathscr{S}(\kappa)$.*

Proof. For every $f \in C_c^+(G)$ we find

$$\infty > \int_0^\infty \langle \rho_\lambda, f\rangle \, d\mu(\lambda) = \int_0^\infty \left(\int_0^\infty \langle \mu_t, f\rangle e^{-\lambda t} dt \right) d\mu(\lambda) = \int_0^\infty \langle \mu_t, f\rangle \mathscr{L}\mu(t)\,dt,$$

and it follows that the completely monotone function $\mathscr{L}\mu$ is integrable over $]0, 1[$. The assertion now follows by Theorem 14.21 using 14.13 and 14.14. □

14.29. Remark. The integral (14) exists for every positive measure μ on $[0, \infty[$ for which $\int_1^\infty \frac{1}{x} d\mu(x) < \infty$.

14.30. Example. Let $(\mu_t^\alpha)_{t>0}$ for $\alpha \in]0, 2]$ be the symmetric stable semigroup of order α on $\mathbb{R}^n$, with associated negative definite function $y \mapsto \|y\|^\alpha$, cf. 10.5.

The function $y \mapsto \|y\|^{-\alpha}$ on $\mathbb{R}^n$ is locally integrable precisely in the following cases

$$n=1:\ 0<\alpha<1,$$
$$n=2:\ 0<\alpha<2,$$
$$n\geqq 3:\ 0<\alpha\leqq 2,$$

and $(\mu_t^\alpha)_{t>0}$ is integrable for these values of n and α. For $n=1$, $\alpha \in [1, 2]$ and $n=2$, $\alpha=2$ $(\mu_t^\alpha)_{t>0}$ is recurrent, cf. Corollary 13.27.

In the integrable cases the potential kernel $\kappa_{n,\alpha}$ for $(\mu_t^\alpha)_{t>0}$ has a density, which by (12) and 10.5 is given for $x \neq 0$ by

$$\frac{1}{\Gamma\left(\frac{\alpha}{2}\right)}\int_0^\infty (4\pi t)^{-\frac{n}{2}}\exp\left(-\frac{\|x\|^2}{4t}\right)t^{\frac{\alpha}{2}-1}\,dt$$

$$=\frac{\Gamma\left(\frac{n-\alpha}{2}\right)}{2^\alpha \pi^{n/2}\Gamma\left(\frac{\alpha}{2}\right)}\|x\|^{\alpha-n},$$

i.e.

$$\kappa_{n,\alpha}=\frac{\Gamma\left(\frac{n-\alpha}{2}\right)}{2^\alpha \pi^{n/2}\Gamma\left(\frac{\alpha}{2}\right)}\|x\|^{\alpha-n}dx. \tag{15}$$

The measure $\kappa_{n,\alpha}$ is called the *Riesz kernel of order* α on $\mathbb{R}^n$. $\Bigl($To verify this formula one makes the substitution $s=-\dfrac{\|x\|^2}{4t}\Bigr)$.

In the case $n \geqq 3$ the Brownian semigroup $(\mu_t)_{t>0}$ is transient, so by Proposition 14.26 we get

$$\kappa_{n,\alpha} * \kappa_{n,\beta} = \kappa_{n,\alpha+\beta} \tag{16}$$

for $\alpha, \beta, \alpha+\beta \in]0,2]$ and $n \geqq 3$.

14.31. Exercise. The formula (16) holds for $n=1$ and $\alpha, \beta, \alpha+\beta \in]0,1[$ and for $n=2$ and $\alpha, \beta, \alpha+\beta \in]0,2[$.

14.32. The convex cones $\mathscr{H}$ and $\mathscr{S}$ are introduced and studied in Hirsch [2] and [3]. The Stieltjes transformation is however classical, cf. Widder [1].

A slightly different version of Corollary 14.9 is proved in Itô [3], cf. Theorem 17.11. The proof of Lemma 14.19 is due to Hirsch [4]. Fractional powers of Hunt kernels have been studied by Faraut [1]. The Propositions 14.27 and 14.28 appear here as special cases of Theorem 14.21 but were originally proved by other methods, cf. Hirsch [1] and Itô [1] and [2].

§ 15. Convergence Lemmas and Potential Theoretic Principles

A convolution kernel (i.e. a positive measure) κ on a LCA-group G defines a mapping $\sigma \mapsto \kappa * \sigma$ of $D^+(\kappa)$ into $M^+(G)$ and the study of this mapping is called potential theory of κ.

Convolution of positive measures is in general not a vaguely continuous operation. It will however be important for the potential theory of κ to know conditions under which convolution is vaguely continuous, and we shall begin with a treatment of some conditions of this kind. Then we discuss certain potential theoretic principles which κ may satisfy e.g. the balayage principle, the complete maximum principle and the principle of unicity of mass. Finally it is proved that an elementary kernel satisfies the balayage principle for all open sets.

15.1. Lemma. *Let $(\mu_\alpha)_{\alpha\in A}$, $(\nu_\alpha)_{\alpha\in A}$ and $(\lambda_\alpha)_{\alpha\in A}$ be vaguely convergent nets on $M^+(G)$ with limits μ, ν and λ. If for all $\alpha\in A$ the convolution $\mu_\alpha*\nu_\alpha$ exists and satisfies $\mu_\alpha*\nu_\alpha\leqq\lambda_\alpha$, then the convolution $\mu*\nu$ exists and satisfies $\mu*\nu\leqq\lambda$.*

Proof. Let $f\in C_c^+(G)$. We shall show that (cf. 1.8)

$$\int f(x+y)\,d\mu\otimes\nu(x,y)\leqq\int f(z)\,d\lambda(z). \tag{1}$$

For a function $g\in C_c^+(G\times G)$ satisfying $g(x,y)\leqq f(x+y)$ for $x,y\in G$ we find

$$\begin{aligned}\int g(x,y)\,d\mu\otimes\nu(x,y)&=\lim_A\int g(x,y)\,d\mu_\alpha\otimes\nu_\alpha(x,y)\\&\leqq\limsup_A\int f(x+y)\,d\mu_\alpha\otimes\nu_\alpha(x,y)\\&\leqq\limsup_A\int f(z)\,d\lambda_\alpha(z)\\&=\int f(z)\,d\lambda(z),\end{aligned}$$

and (1) follows by taking supremum over all g as above. □

15.2. Exercise. Let $(\mu_\alpha)_{\alpha\in A}$ and $(\nu_\alpha)_{\alpha\in A}$ be vaguely convergent nets on $M^+(G)$ with limits μ and ν. If there exists a compact set $K\subseteq G$ such that $\operatorname{supp}(\nu_\alpha)\subseteq K$ for all $\alpha\in A$, then the convolutions $\mu_\alpha*\nu_\alpha$ and $\mu*\nu$ exist for all $\alpha\in A$, and the net $(\mu_\alpha*\nu_\alpha)_{\alpha\in A}$ converges vaguely to $\mu*\nu$.

15.3. Lemma. *Let $(\mu_\alpha)_{\alpha\in A}$ and $(\nu_\alpha)_{\alpha\in A}$ be monotonically increasing nets on $M^+(G)$ with vague limits μ and ν. If the convolution $\mu_\alpha*\nu_\alpha$ exists for all $\alpha\in A$, and there is a $\lambda\in M^+(G)$ such that $\mu_\alpha*\nu_\alpha\leqq\lambda$ for all $\alpha\in A$, then the convolution $\mu*\nu$ exists, and $\mu*\nu$ is vague limit of the net $(\mu_\alpha*\nu_\alpha)_{\alpha\in A}$.*

Proof. Since the net $(\mu_\alpha*\nu_\alpha)_{\alpha\in A}$ is increasing and bounded, it is vaguely convergent. We put

$$\lambda_0=\lim_A\mu_\alpha*\nu_\alpha.$$

By Lemma 15.1, the convolution $\mu*\nu$ exists and $\mu*\nu\leqq\lambda_0$, and since $\mu_\alpha*\nu_\alpha\leqq\mu*\nu$ for all $\alpha\in A$ we have $\lambda_0\leqq\mu*\nu$. □

15.4. Exercise. Let $(\mu_\alpha)_{\alpha\in A}$ and $(\nu_\alpha)_{\alpha\in A}$ be monotonically decreasing nets on $M^+(G)$ with vague limits μ and ν. If the convolution $\mu_\alpha*\nu_\alpha$ exists for all $\alpha\in A$, then the convolution $\mu*\nu$ exists, and $\mu*\nu$ is the vague limit of the net $(\mu_\alpha*\nu_\alpha)_{\alpha\in A}$.

15.5. Proposition (Deny [5]). *Let $(\mu_\alpha)_{\alpha\in A}$ be a vaguely convergent net on $M^+(G)$ with limit μ, and let $\nu,\lambda\in M^+(G)$ be such that $\lambda\neq0$ and the convolution $\mu_\alpha*\lambda$ exists and satisfies $\mu_\alpha*\lambda\leqq\nu$ for all $\alpha\in A$. Let $\rho\in M^+(G)$ be such that the convolution $\nu*\rho$ exists. Then the convolutions $\mu_\alpha*\rho$ and $\mu*\rho$ exist, and the net $(\mu_\alpha*\rho)_{\alpha\in A}$ is vaguely convergent with limit $\mu*\rho$.*

Proof. By Lemma 15.1 the convolution $\mu*\lambda$ exists and satisfies $\mu*\lambda\leqq\nu$. Therefore the convolution of $\mu*\lambda$ and ρ exists, and since $\lambda\neq 0$ the convolution $\mu*\rho$ exists, cf. 1.10. In the same manner we see that the convolution $\mu_\alpha*\rho$ exists for all $\alpha\in A$.

Let $f\in C_c^+(G)$. For any $g\in C_c^+(G)$ such that $g\leqq\check{\rho}*f$ we find

$$\begin{aligned}\langle\mu, g\rangle &= \lim_A \langle\mu_\alpha, g\rangle\\ &\leqq \liminf_A \langle\mu_\alpha, \check{\rho}*f\rangle\\ &= \liminf_A \langle\mu_\alpha*\rho, f\rangle,\end{aligned}$$

and this implies that

$$\langle\mu*\rho, f\rangle = \langle\mu, \check{\rho}*f\rangle \leqq \liminf_A \langle\mu_\alpha*\rho, f\rangle.$$

The assertion will be proved when we have seen that

$$\limsup_A \langle\mu_\alpha*\rho, f\rangle \leqq \langle\mu*\rho, f\rangle. \tag{2}$$

By Lemma 1.9 we choose $h\in C_c^+(G)$ such that $f\leqq\check{\lambda}*h$. For a compact set $K\subseteq G$ we denote by ρ_K the restriction of ρ to K and we put $\rho'_K=\rho-\rho_K$. As K increases to G the measure ρ'_K decreases and converges vaguely to 0. It follows by Exercise 15.4 that there exists for a given $\varepsilon>0$ a compact set $K\subseteq G$ such that $\langle\nu*\rho'_K, h\rangle\leqq\varepsilon$, and hence

$$\begin{aligned}\limsup_A \langle\mu_\alpha*\rho, f\rangle &\leqq \limsup_A \langle\mu_\alpha*\rho_K, f\rangle + \limsup_A \langle\mu_\alpha*\rho'_K, f\rangle\\ &\leqq \langle\mu*\rho_K, f\rangle + \varepsilon,\end{aligned}$$

because ρ_K has compact support (Exercise 15.2) and

$$\begin{aligned}\langle\mu_\alpha*\rho'_K, f\rangle &\leqq \langle\mu_\alpha*\rho'_K, \check{\lambda}*h\rangle \leqq \langle\mu_\alpha*\lambda*\rho'_K, h\rangle\\ &\leqq \langle\nu*\rho'_K, h\rangle \leqq \varepsilon,\end{aligned}$$

and the inequality (2) follows. □

15.6. Corollary. *Let $(\mu_\alpha)_{\alpha\in A}$ be a vaguely convergent net on $M^+(G)$ with limit μ and let $\nu\in M^+(G)$ be such that $\mu_\alpha\leqq\nu$ for all $\alpha\in A$. For every measure $\rho\in D^+(\nu)$ the net $(\mu_\alpha*\rho)_{\alpha\in A}$ converges vaguely to $\mu*\rho$.*

Proof. Put $\lambda=\varepsilon_0$ in Proposition 15.5. □

15.7. Lemma. *Let $(\mu_\alpha)_{\alpha\in A}$ be a family of positive measures on G and let $\kappa\in M^+(G)$ be a non-zero measure such that the convolution $\kappa*\mu_\alpha$ exists for all $\alpha\in A$ and the family $(\kappa*\mu_\alpha)_{\alpha\in A}$ is vaguely bounded. Then the family $(\mu_\alpha)_{\alpha\in A}$ is vaguely bounded.*

Proof. Let $f \in C_c^+(G)$ and choose (by 1.9) $h \in C_c^+(G)$ such that $f \leqq \check{\kappa} * h$. For $\alpha \in A$ we have

$$\langle \mu_\alpha, f \rangle \leqq \langle \mu_\alpha, \check{\kappa} * h \rangle = \langle \kappa * \mu_\alpha, h \rangle,$$

and since $(\kappa * \mu_\alpha)_{\alpha \in A}$ is vaguely bounded, this implies

$$\sup_A \langle \mu_\alpha, f \rangle < \infty. \quad \square$$

We shall now introduce some of the fundamental potential theoretic principles for convolution kernels (i.e. positive measures) on G.

15.8. Definition. Let κ be a convolution kernel on G. For $\mu \in D^+(\kappa)$ and an open set $\omega \subseteq G$ a measure $\mu' \in D^+(\kappa)$ is said to be a *κ-balayaged measure of μ on ω*, if it satisfies

(i) $\operatorname{supp}(\mu') \subseteq \bar{\omega}$.
(ii) $\kappa * \mu' \leqq \kappa * \mu$ as measures on G.
(iii) The restrictions of the measures $\kappa * \mu'$ and $\kappa * \mu$ to ω are equal.

15.9. Definition. A convolution kernel κ is said to satisfy the *balayage principle*, if there exists for every $\mu \in M_c^+(G)$ and every open relatively compact set $\omega \subseteq G$ a κ-balayaged measure of μ on ω.

And κ is said to satisfy the *balayage principle for all open sets*, if there exists for every $\mu \in M_c^+(G)$ and every open set $\omega \subseteq G$ a κ-balayaged measure of μ on ω.

15.10. Definition. A convolution kernel κ is said to satisfy the *domination principle*, if for all $f, g \in C_c^+(G)$ we have the implication:

$$[\forall x \in \operatorname{supp}(f)\colon \kappa * f(x) \leqq \kappa * g(x)] \Rightarrow [\forall x \in G\colon \kappa * f(x) \leqq \kappa * g(x)].$$

And κ is said to satisfy the *complete maximum principle*, if for all $f, g \in C_c^+(G)$ and all constants $a \geqq 0$ we have the following implication:

$$[\forall x \in \operatorname{supp}(f)\colon \kappa * f(x) \leqq \kappa * g(x) + a] \Rightarrow [\forall x \in G\colon \kappa * f(x) \leqq \kappa * g(x) + a].$$

15.11. Definition. A convolution kernel κ is said to satisfy the *principle of unicity of mass*, if for all $\sigma_1, \sigma_2 \in D^+(\kappa)$ we have the implication:

$$\kappa * \sigma_1 = \kappa * \sigma_2 \Rightarrow \sigma_1 = \sigma_2.$$

15.12. Remark. It is clear that if the convolution kernel κ satisfies one of the above principles (15.9–15.11), then the reflected convolution kernel $\check{\kappa}$ satisfies the same principle.

We refer to Choquet and Deny [3] for a detailed discussion of these and other potential theoretic principles.

The following result gives an example of the "many" relations which link these potential theoretic principles.

15.13. Theorem. *A convolution kernel κ satisfies the balayage principle if and only if it satisfies the domination principle.*

Proof. Suppose first that κ satisfies the balayage principle, and let $f, g \in C_c^+(G)$ be such that

$$\kappa * f(x) \leqq \kappa * g(x) \qquad \text{for all } x \in \operatorname{supp}(f).$$

Let $x \in G$ be arbitrary and denote by ε'_x a $\check{\kappa}$-balayaged measure, cf. 15.12, of ε_x on the open relatively compact set $\{f>0\}$.

Then we have

$$\begin{aligned} \kappa * f(x) &= \langle \varepsilon_x, \kappa * f \rangle = \langle \check{\kappa} * \varepsilon_x, f \rangle = \langle \check{\kappa} * \varepsilon'_x, f \rangle \\ &= \langle \varepsilon'_x, \kappa * f \rangle \leqq \langle \varepsilon'_x, \kappa * g \rangle = \langle \check{\kappa} * \varepsilon'_x, g \rangle \\ &\leqq \langle \check{\kappa} * \varepsilon_x, g \rangle = \langle \varepsilon_x, \kappa * g \rangle = \kappa * g(x), \end{aligned}$$

which shows that κ satisfies the domination principle.

For a closed set $F \subseteq G$ we put

$$M^+(F) = \{\mu \in M^+(G) \mid \operatorname{supp}(\mu) \subseteq F\}.$$

The convolution kernel κ satisfies the balayage principle if and only if

$$\kappa * (M_c^+(G)) \subseteq \kappa * (M^+(\bar{\omega})) + M^+(\complement \omega) \tag{3}$$

for every open, relatively compact set $\omega \subseteq G$.

Suppose now that κ satisfies the domination principle and that $\kappa \neq 0$, and let $\omega \subseteq G$ be an open, relatively compact set. We shall see that the inclusion (3) holds.

The set $M^+(\bar{\omega}) \cap M_1^+(G)$ is vaguely compact and by the continuity of the mapping $\mu \mapsto \kappa * \mu$ of $M^+(\bar{\omega})$ into $M^+(G)$ (with the vague topologies), cf. 15.2, it follows that the set

$$A = \{\kappa * \mu \mid \mu \in M^+(\bar{\omega}) \cap M_1^+(G)\}$$

is vaguely compact. Since $0 \notin A$ the cone generated by A

$$\bigcup_{\lambda \geqq 0} \lambda A = \kappa * (M^+(\bar{\omega})),$$

is vaguely closed. The sum of two vaguely closed sets of positive measures is again vaguely closed (Bourbaki [1] p. 60), and therefore

$$C = \kappa * (M^+(\bar{\omega})) + M^+(\complement \omega)$$

is a vaguely closed convex cone of positive measures.

For a positive measure τ on G such that $\tau \notin C$ there exists by the Hahn-Banach theorem a real function $f \in C_c(G)$ such that

$$\langle \tau, f \rangle < 0 \leqq \langle \sigma, f \rangle \quad \text{for all } \sigma \in C. \tag{4}$$

In particular

$$\langle \kappa * \varepsilon_x, f \rangle = \check{\kappa} * f(x) \geqq 0 \quad \text{for all } x \in \overline{\omega},$$

and

$$\langle \varepsilon_x, f \rangle = f(x) \geqq 0 \quad \text{for all } x \in \complement \omega.$$

Putting $f^+ = \sup(f, 0)$ and $f^- = \sup(-f, 0)$ we then have

$$\check{\kappa} * f^-(x) \leqq \check{\kappa} * f^+(x) \quad \text{for } x \in \overline{\omega}$$

and

$$\operatorname{supp}(f^-) \subseteq \overline{\omega},$$

so by the domination principle for $\check{\kappa}$, cf. 15.12, we get that $\check{\kappa} * f \geqq 0$.

This implies that

$$\langle \kappa * \mu, f \rangle \geqq 0 \quad \text{for all } \mu \in M_c^+(G),$$

which shows that τ does not belong to $\kappa * (M_c^+(G))$, hence that (3) holds. □

As an application of Theorem 15.13 we prove the following result, cf. Deny [1] p. 97.

15.14. Proposition. *The set of convolution kernels on G satisfying the balayage principle is vaguely closed.*

Proof. Let $(\kappa_\alpha)_{\alpha \in A}$ be a net of convolution kernels satisfying the balayage principle and suppose that

$$\lim_A \kappa_\alpha = \kappa \text{ vaguely.}$$

By Theorem 15.13 it suffices to prove that κ satisfies the domination principle, and we may suppose that $\kappa \neq 0$. Let $f, g \in C_c^+(G)$ be such that

$$\kappa * f(x) \leqq \kappa * g(x) \quad \text{for all } x \in \operatorname{supp}(f),$$

and choose a function $h \in C_c^+(G)$ such that $\kappa * h(x) > 0$ for all $x \in \operatorname{supp}(f)$, cf. 1.9. For fixed $a > 0$ we then have

$$\kappa * f(x) < \kappa * (g + ah)(x) \quad \text{for } x \in \operatorname{supp}(f),$$

and it follows by Exercise 1.20 that there exists an $\alpha_0 \in A$ such that for all $\alpha \in A$ with $\alpha \geqq \alpha_0$

$$\kappa_\alpha * f(x) < \kappa_\alpha * (g + ah)(x) \quad \text{for } x \in \operatorname{supp}(f).$$

By the domination principle for κ_α we get

$$\kappa_\alpha * f(x) \leqq \kappa_\alpha * (g + ah)(x) \quad \text{for } x \in G \text{ and } \alpha \geqq \alpha_0,$$

and going to the limit for $\alpha \in A$ we find

$$\kappa * f(x) \leqq \kappa * (g + ah)(x) \quad \text{for } x \in G.$$

The constant $a > 0$ being arbitrary we finally get

$$\kappa * f(x) \leqq \kappa * g(x) \quad \text{for all } x \in G. \quad \square$$

Let μ be a positive measure on G such that $\mu(G) \leqq 1$.

15.15. Definition. A positive measure ξ on G is called *μ-superharmonic* (resp. *μ-harmonic*) if $\xi \in D^+(\mu)$ and $\mu * \xi \leqq \xi$ (resp. $\mu * \xi = \xi$).

The set $S(\mu)$ of μ-superharmonic measures is a convex cone, which by Lemma 15.1 is vaguely closed, and the set $H(\mu)$ of μ-harmonic measures is a subcone of $S(\mu)$.

The Haar measure ω_G of G is μ-superharmonic, and it is μ-harmonic if and only if $\mu(G) = 1$. This follows from the formula

$$\mu * \omega_G = \mu(G)\,\omega_G.$$

15.16. Proposition. (i) *The infimum of an arbitrary set of μ-superharmonic measures is μ-superharmonic.*

(ii) *The supremum of an upward filtering and bounded set of μ-superharmonic (resp. μ-harmonic) measures is μ-superharmonic (resp. μ-harmonic).*

(iii) *Let ξ be a μ-superharmonic (resp. μ-harmonic) measure and let $\sigma \in D^+(\xi)$. Then $\xi * \sigma$ is μ-superharmonic (resp. μ-harmonic).*

Proof. (i) and (iii) are obvious, and (ii) follows by application of Lemma 15.3. $\square$

For an integral representation of the elements of $H(\mu)$ see Choquet and Deny [3] and Deny [3].

15.17. Lemma. *The set $S(\mu) \setminus H(\mu)$ is non-empty if and only if the series*

$$\sum_{n=0}^{\infty} \mu^n$$

is vaguely convergent. In the affirmative case the measure

$$\kappa = \sum_{n=0}^{\infty} \mu^n$$

is in $S(\mu)\setminus H(\mu)$.

Proof. Suppose that $\xi \in S(\mu)\setminus H(\mu)$. Then $\sigma = \xi - \mu * \xi$ is a positive non-zero measure, and for $N \in \mathbb{N}$ we have

$$\left(\sum_{n=0}^{N} \mu^n\right) * \sigma = \xi - \mu^{N+1} * \xi \leqq \xi.$$

By Lemma 15.7 the increasing sequence $(\sum_{n=0}^{N} \mu^n)_{N\in\mathbb{N}}$ is bounded and therefore vaguely convergent.

Suppose next that the series $\sum_{n=0}^{\infty} \mu^n$ is vaguely convergent with sum κ. Then

$$\mu * \kappa = \kappa - \varepsilon_0,$$

which shows that $\kappa \in S(\mu)\setminus H(\mu)$. □

Suppose now that the series $\sum_{n=0}^{\infty} \mu^n$ is vaguely convergent. The convolution kernel $\kappa = \sum_{n=0}^{\infty} \mu^n$ is the elementary kernel determined by μ, cf. 13.6.

For every $\sigma \in D^+(\kappa)$ the measure $\kappa * \sigma$ is called the *κ-potential generated* by σ, and it is a μ-superharmonic measure. It is μ-harmonic if and only if $\sigma = 0$.

The next result is a special case of the *Riesz decomposition theorem*, which will be proved later, cf. 16.7.

15.18. Proposition. *Let κ be the elementary kernel determined by μ. Every μ-superharmonic measure ξ can be written*

$$\xi = \kappa * \sigma + \eta,$$

where $\sigma \in D^+(\kappa)$ and η is μ-harmonic.

In such a decomposition σ and η are uniquely determined and given by the formulas

$$\sigma = \xi - \mu * \xi, \qquad \eta = \lim_{n\to\infty} \mu^n * \xi \ \textit{vaguely}.$$

The measures $\kappa * \sigma$ and η are called respectively the potential part *and the* μ-harmonic part *of ξ.*

Proof. Suppose first that the μ-superharmonic measure ξ has a decomposition

$$\xi = \kappa * \sigma + \eta,$$

where $\sigma \in D^+(\kappa)$ and η is μ-harmonic. Since

$$\mu * \xi = \mu * (\kappa * \sigma) + \eta = (\mu * \kappa) * \sigma + \eta$$

we get

$$\xi - \mu * \xi = (\kappa - \mu * \kappa) * \sigma = \sigma .$$

Furthermore, for $n \in \mathbb{N}$ we have

$$\mu^n * \xi = (\mu^n * \kappa) * \sigma + \eta ,$$

and since

$$\mu^n * \kappa = \sum_{k=n}^{\infty} \mu^k$$

decreases to 0 as n tends to ∞, we get by Exercise 15.4 that

$$\eta = \lim_{n \to \infty} \mu^n * \xi \text{ vaguely.}$$

This proves the uniqueness of σ and η.

Let ξ be a μ-superharmonic measure. The sequence $(\mu^n * \xi)_{n \in \mathbb{N}}$ being decreasing, the vague limit

$$\eta = \lim_{n \to \infty} \mu^n * \xi$$

exists, and it follows by Exercise 15.4 that η is μ-harmonic. The measure $\sigma = \xi - \mu * \xi$ is a positive measure, and we find for $N \in \mathbb{N}$

$$\left(\sum_{n=0}^{N} \mu^n \right) * \sigma = \xi - \mu^{N+1} * \xi ;$$

letting $N \to \infty$ we conclude by Lemma 15.3 that $\kappa * \sigma = \xi - \eta$. □

15.19. Corollary. *An elementary kernel satisfies the principle of unicity of mass.*

15.20. Corollary. *Any μ-superharmonic measure ξ which is dominated by a potential is a potential, i.e. there exists $\tau \in D^+(\kappa)$ such that $\xi = \kappa * \tau$.*

Proof. Let $\xi = \kappa * \tau + \eta$ be the decomposition of ξ from Proposition 15.18. By assumption there exists $\sigma \in D^+(\kappa)$ such that $\xi \leqq \kappa * \sigma$, and for the μ-harmonic part η of ξ we then find

$$0 \leqq \eta = \lim_{n \to \infty} \mu^n * \xi \leqq \lim_{n \to \infty} \mu^n * (\kappa * \sigma) = 0 . \quad \square$$

15.21. Proposition. *A positive measure κ on G is potential kernel for at most one transient convolution semigroup.*

Proof. Suppose that $(\mu_t')_{t>0}$ and $(\mu_t'')_{t>0}$ are transient convolution semigroups with κ as potential kernel. Denoting by $(\rho_\lambda')_{\lambda>0}$ and $(\rho_\lambda'')_{\lambda>0}$ the resolvents for

$(\mu'_t)_{t>0}$ and $(\mu''_t)_{t>0}$ we get by Proposition 13.3 that

$$\kappa=(\lambda\kappa+\varepsilon_0)*\rho'_\lambda=(\lambda\kappa+\varepsilon_0)*\rho''_\lambda \quad \text{for } \lambda>0.$$

By Proposition 13.7 the measure $\lambda\kappa+\varepsilon_0$ is an elementary kernel, so by Corollary 15.19 it follows that $\rho'_\lambda=\rho''_\lambda$ for all $\lambda>0$. This in turn implies that $\mu'_t=\mu''_t$ for all $t>0$, cf. 8.12. □

15.22. Proposition. *Let μ and ν be positive measures on G such that $\mu(G)\leqq 1$ and $\nu(G)\leqq 1$ and suppose that the elementary kernels $\sum_{n=0}^{\infty}\mu^n$ and $\sum_{n=0}^{\infty}\nu^n$ determined by μ and ν exist. If*

$$\sum_{n=0}^{\infty}\mu^n=\sum_{n=0}^{\infty}\nu^n$$

then $\mu=\nu$.

Proof. For the elementary kernel

$$\kappa=\sum_{n=0}^{\infty}\mu^n=\sum_{n=0}^{\infty}\nu^n$$

we have

$$\kappa*\mu=\sum_{n=1}^{\infty}\mu^n=\kappa-\varepsilon_0=\sum_{n=1}^{\infty}\nu^n=\kappa*\nu,$$

and it follows by Corollary 15.19 that $\mu=\nu$. □

15.23. Proposition. *An elementary kernel $\kappa=\sum_{n=0}^{\infty}\mu^n$ satisfies the balayage principle for all open sets.*

Proof. Let $\sigma\in D^+(\kappa)$ and let $\omega\subseteq G$ be an open set. The infimum of all μ-superharmonic measures dominating $\kappa*\sigma$ on ω is a μ-superharmonic measure ξ_0 such that $\xi_0=\kappa*\sigma$ on ω and $\xi_0\leqq\kappa*\sigma$. By Corollary 15.20 there exists $\sigma'\in D^+(\kappa)$ such that $\xi_0=\kappa*\sigma'$.

The measure σ' satisfies (ii) and (iii) of Definition 15.8, and in order to prove that $\operatorname{supp}(\sigma')\subseteq\overline{\omega}$ we denote by σ'_1 (resp. σ'_2) the restriction of σ' to ω (resp. $\complement\omega$). Then we find

$$\kappa*(\sigma'_1+\mu*\sigma'_2)=\kappa*\sigma'_1+(\kappa-\varepsilon_0)*\sigma'_2=\kappa*\sigma'-\sigma'_2,$$

and since σ'_2 is zero on ω, the μ-superharmonic measure $\kappa*(\sigma'_1+\mu*\sigma'_2)$ is equal to $\kappa*\sigma'$ and hence to $\kappa*\sigma$ on ω. It follows that

$$\kappa*(\sigma'_1+\mu*\sigma'_2)\geqq\kappa*\sigma',$$

and then $\sigma'_2=0$, which implies that $\sigma'=\sigma'_1$. □

15.24. Remark. The set of convolution kernels on G satisfying the balayage principle is clearly a cone but in general not a convex cone. This follows e.g.

by the fact that the support of a convolution kernel satisfying the balayage principle is a semigroup in G, cf. Choquet and Deny [3], and the counter-example of 13.8 can be applied. See also Exercise 15.25 below.

15.25. Exercise. Each of the convolution kernels $\kappa_1=\sum_{n=0}^{\infty}\varepsilon_1^n$ and $\kappa_2=1_{]0,\infty[}(x)dx$ on $G=\mathbb{R}$ satisfies the domination principle, but $\kappa_1+\kappa_2$ does not satisfy the domination principle.

15.26. The elementary kernels have been introduced by Deny [1]. See also Deny [2] and [5] for a discussion of elementary kernels in a more general (not necessarily translation invariant) context.

The theory of elementary kernels will not be developed further, because it is a special case of the potential theory for transient convolution semigroups, which will be the topic of the next paragraph.

The importance of the elementary kernels lies in the fact, cf. 13.7, that the potential kernel for a transient convolution semigroup is vague limit of elementary kernels. This will allow us to show that every such potential kernel satisfies the balayage principle for all open sets.

§ 16. Excessive Measures

For a transient convolution semigroup $(\mu_t)_{t>0}$ with potential kernel κ on the LCA-group G, measures of the form $\kappa * \sigma$, where $\sigma \in D^+(\kappa)$, will be called κ-potentials, and in the study of such measures, it is natural (cf. 16.29) to consider the larger class of measures, which are excessive with respect to $(\mu_t)_{t>0}$.

The first main result is the Riesz decomposition theorem for excessive measures. Then it is shown that the potential kernel satisfies the balayage principle for all open sets and the balayage is related to the notion of the reduced measure of an excessive measure over an open set.

The study of the reduced measure leads to the equilibrium principle and the complete maximum principle for the potential kernel.

Let $(\mu_t)_{t>0}$ be a convolution semigroup on G.

16.1. Definition. A positive measure ξ on G is said to be *excessive* (resp. *invariant*) with respect to $(\mu_t)_{t>0}$ if for all $t>0$, ξ is μ_t-superharmonic (resp. μ_t-harharmonic), i.e. if for all $t>0$, $\xi \in D^+(\mu_t)$ and $\mu_t * \xi \leq \xi$ (resp. $\mu_t * \xi = \xi$).

The set of excessive measures $\bigcap_{t>0} S(\mu_t)$ is a vaguely closed convex cone and the set of invariant measures $\bigcap_{t>0} H(\mu_t)$ is a convex subcone of the cone of excessive measures.

The Haar measure ω_G on G is excessive, and it is invariant if and only if all the measures μ_t for $t>0$ are probability measures.

Proposition 15.16 gives the following result for excessive measures.

16.2. Proposition. (i) *The infimum of an arbitrary set of excessive measures is excessive.*

(ii) *The supremum of an upward filtering and bounded set of excessive (resp. invariant) measures is excessive (resp. invariant).*

(iii) *Let ξ be an excessive (resp. invariant) measure and let $\sigma \in D^+(\xi)$. Then $\xi * \sigma$ is excessive (resp. invariant).*

16.3. Proposition. *Let ξ be an excessive measure. The mapping (putting $\mu_0 = \varepsilon_0$), $t \mapsto \mu_t * \xi$ of $[0, \infty[$ into $M^+(G)$ is decreasing and vaguely continuous from the right. In particular*

$$\xi = \lim_{t \to 0} \mu_t * \xi \quad \textit{vaguely}.$$

Proof. For $s, t > 0$ we have

$$\mu_{t+s} * \xi = \mu_t * (\mu_s * \xi) \leqq \mu_t * \xi,$$

which shows that the mapping $t \mapsto \mu_t * \xi$ is decreasing. For $f \in C_c^+(G)$ we find

$$\langle \xi, f \rangle \leqq \liminf_{t \to 0} \langle \mu_t * \xi, f \rangle = \lim_{t \to 0} \langle \mu_t * \xi, f \rangle,$$

but since

$$\langle \mu_t * \xi, f \rangle \leqq \langle \xi, f \rangle \qquad \text{for all } t > 0,$$

we get

$$\xi = \lim_{t \to 0} \mu_t * \xi \text{ vaguely}.$$

For $t_0 > 0$ the measure $\mu_{t_0} * \xi$ is excessive, and it follows that

$$\mu_{t_0} * \xi = \lim_{t \to 0} \mu_t * (\mu_{t_0} * \xi) = \lim_{s \searrow t_0} \mu_s * \xi \quad \text{vaguely.} \quad \square$$

16.4. Exercise. Let $(\mu_t)_{t>0}$ be a convolution semi-group on G and let $(\rho_\lambda)_{\lambda>0}$ be the resolvent for $(\mu_t)_{t>0}$.

1) A positive measure ξ is excessive (resp. invariant) with respect to $(\mu_t)_{t>0}$ if and only if $\xi \in D^+(\rho_\lambda)$ and $\lambda \rho_\lambda * \xi \leqq \xi$ (resp. $\lambda \rho_\lambda * \xi = \xi$) for all $\lambda > 0$.

2) For an excessive measure ξ the mapping $\lambda \mapsto \lambda \rho_\lambda * \xi$ of $]0, \infty[$ into $M^+(G)$ is increasing and $\lim_{\lambda \to \infty} \lambda \rho_\lambda * \xi = \xi$ vaguely.

3) Let ξ be an excessive measure and suppose that there exists a $t_0 > 0$ such that $\mu_{t_0} * \xi = \xi$, or a $\lambda_0 > 0$ such that $\lambda_0 \rho_{\lambda_0} * \xi = \xi$. Then ξ is invariant.

16.5. From now on we consider a fixed transient convolution semigroup $(\mu_t)_{t>0}$ on G with potential kernel κ and resolvent $(\rho_\lambda)_{\lambda>0}$. The notion of excessiveness depends "only" on κ, since a (transient) convolution semigroup is uniquely determined by its potential kernel (cf. 15.21).

For $t > 0$ we have

$$\mu_t * \kappa = \int_t^\infty \mu_s \, ds,$$

which shows that κ is excessive but not invariant.

For $\sigma \in D^+(\kappa)$ the measure $\kappa * \sigma$ is called the *κ-potential generated* by σ. The potential $\kappa * \sigma$ is excessive, and invariant if and only if $\sigma = 0$.

16.6. Lemma. *For $\sigma \in D^+(\kappa)$ and $t>0$ the convolution $\mu_t * \sigma$ exists. Putting $\mu_0 = \varepsilon_0$, the mapping $t \mapsto \mu_t * \sigma$ of $[0, \infty[$ into $M^+(G)$ is vaguely continuous and*

$$\kappa * \sigma = \int_0^\infty \mu_t * \sigma \, dt \quad \textit{vaguely}.$$

Proof. We have that $\mu_t * \kappa \leqq \kappa$ for all $t>0$. Since $\sigma \in D^+(\kappa)$ and $\kappa \neq 0$ the continuity assertion of the lemma is an immediate consequence of Proposition 15.5, and the above formula follows easily. □

The next theorem, which is due to Deny [5], is called the *Riesz decomposition theorem* on account of a theorem of F. Riesz in classical potential theory, cf. Brelot [1] p. 47.

16.7. Theorem. *Every excessive measure ξ can be written*

$$\xi = \kappa * \sigma + \eta,$$

where $\sigma \in D^+(\kappa)$ and η is an invariant measure.

In such a decomposition σ and η are uniquely determined and given by the formulas

$$\sigma = \lim_{t \to 0} \frac{1}{t}(\xi - \mu_t * \xi)$$

and

$$\eta = \lim_{t \to \infty} \mu_t * \xi,$$

the limits being taken in the vague topology.

The measures $\kappa * \sigma$ and η are called respectively the potential part *of ξ and the* invariant part *of ξ, and the decomposition of ξ as sum of the potential part of ξ and the invariant part of ξ is called the* Riesz decomposition *of ξ.*

Proof. Suppose first that the excessive measure ξ can be written

$$\xi = \kappa * \sigma + \eta,$$

where $\sigma \in D^+(\kappa)$ and η is an invariant measure. Let $t>0$. Since

$$\mu_t * (\kappa * \sigma) = (\mu_t * \kappa) * \sigma \quad \text{and} \quad \mu_t * \eta = \eta,$$

we find

$$\xi - \mu_t * \xi = (\kappa - \mu_t * \kappa) * \sigma = \left(\int_0^t \mu_s \, ds \right) * \sigma = \int_0^t \mu_s * \sigma \, ds$$

and

$$\eta = \mu_t * \xi - \int_t^\infty \mu_s * \sigma \, ds.$$

By Lemma 16.6 it follows that

$$\sigma=\lim_{t\to 0}\frac{1}{t}(\xi-\mu_t*\xi) \quad\text{and}\quad \eta=\lim_{t\to\infty}\mu_t*\xi,$$

and this shows that σ and η are uniquely determined.

Let now ξ be an excessive measure.

The mapping $t\mapsto\mu_t*\xi$ being decreasing, the limit

$$\eta=\lim_{t\to\infty}\mu_t*\xi$$

exists, and by Exercise 15.4 we get for every $s>0$

$$\mu_s*\eta=\lim_{t\to\infty}\mu_s*(\mu_t*\xi)=\lim_{t\to\infty}\mu_{t+s}*\xi=\eta,$$

which shows that η is invariant.

For $t>0$ and $n\in\mathbb{N}$ we define

$$\sigma_t=\frac{1}{t}(\xi-\mu_t*\xi) \quad\text{and}\quad \kappa_n=\int_0^n\mu_s\,ds.$$

We first remark that the convolution $\kappa_n*\xi$ exists, because for $f\in C_c^+(G)$ we have

$$\langle\kappa_n,\check{\xi}*f\rangle=\int_0^n\langle\mu_s,\check{\xi}*f\rangle\,ds\leqq\int_0^n\langle\xi,f\rangle\,ds<\infty.$$

Since $\sigma_t\leqq\frac{1}{t}\xi$, the convolution $\kappa_n*\sigma_t$ exists and for $n>t$ we find

$$\begin{aligned}\kappa_n*\sigma_t&=\frac{1}{t}(\kappa_n*\xi-\kappa_n*(\mu_t*\xi))\\&=\frac{1}{t}\left(\int_0^n\mu_s*\xi\,ds-\int_0^n\mu_s*(\mu_t*\xi)\,ds\right)\\&=\frac{1}{t}\left(\int_0^t\mu_s*\xi\,ds-\int_n^{n+t}\mu_s*\xi\,ds\right).\end{aligned}\tag{1}$$

Using that $\eta\leqq\mu_s*\xi\leqq\xi$ for all $s>0$, it follows that for $n>t$

$$\kappa_n*\sigma_t\leqq\xi-\eta.\tag{2}$$

The measures κ_n increase to κ as $n\to\infty$, and by Lemma 15.3 we get that the convolution of κ and σ_t exists for all $t>0$, and that

$$\kappa*\sigma_t\leqq\xi-\eta.$$

By Lemma 15.7 it follows that the family of measures $(\sigma_t)_{t>0}$ is vaguely bounded and hence vaguely relatively compact. Let σ be a vague accumulation point for $(\sigma_t)_{t>0}$ as t tends to 0. There exists a net $(t_i)_{i\in I}$ of positive numbers tending to 0 such that

$$\lim_I \sigma_{t_i} = \sigma.$$

Since the convolution of κ_n and $\xi - \eta$ exists, an application of Proposition 15.5 to (2) gives that

$$\lim_I \kappa_n * \sigma_{t_i} = \kappa_n * \sigma.$$

Using Proposition 16.3 we get

$$\lim_I \frac{1}{t_i} \left(\int_0^{t_i} \mu_s * \xi \, ds - \int_n^{n+t_i} \mu_s * \xi \, ds \right) = \xi - \mu_n * \xi,$$

so by (1) we have

$$\kappa_n * \sigma = \xi - \mu_n * \xi \qquad \text{for all } n \in \mathbb{N}.$$

Since $\kappa_n * \sigma \leqq \xi$, Lemma 15.3 gives that $\sigma \in D^+(\kappa)$ and that

$$\kappa * \sigma = \lim_{n\to\infty} \kappa_n * \sigma = \xi - \lim_{n\to\infty} \mu_n * \xi = \xi - \eta,$$

which shows that

$$\xi = \kappa * \sigma + \eta. \quad \square$$

16.8. Corollary. *The potential kernel κ satisfies the principle of unicity of mass, i.e. for all $\sigma_1, \sigma_2 \in D^+(\kappa)$ we have the implication:*

$$\kappa * \sigma_1 = \kappa * \sigma_2 \Rightarrow \sigma_1 = \sigma_2.$$

16.9. Corollary. *Every excessive measure which is dominated by a potential is a potential.*

Proof. Let $\xi = \kappa * \tau + \eta$ be the Riesz decomposition of the excessive measure ξ. By assumption there exists a measure $\sigma \in D^+(\kappa)$ such that $\xi \leqq \kappa * \sigma$ and for the invariant part η of ξ we then find

$$0 \leqq \eta = \lim_{t\to\infty} \mu_t * \xi \leqq \lim_{t\to\infty} \mu_t * (\kappa * \sigma) = 0. \quad \square$$

16.10. Theorem. *Let $(\xi_\alpha)_{\alpha\in A}$ be a vaguely convergent net of potentials $\xi_\alpha = \kappa * \sigma_\alpha$ with limit ξ, and suppose that there exists a potential $\kappa * \tau$ such that $\xi_\alpha \leqq \kappa * \tau$ for all $\alpha \in A$.*

Then ξ is a potential, $\xi = \kappa * \sigma$, and σ is the vague limit of the net $(\sigma_\alpha)_{\alpha\in A}$.

Proof. The measure ξ is excessive and $\xi \leqq \kappa * \tau$. By Corollary 16.9 there exists a measure $\sigma \in D^+(\kappa)$ such that $\xi = \kappa * \sigma$.

By Corollary 15.6 we get for every $t>0$

$$\lim_A \mu_t * \xi_\alpha = \mu_t * \xi,$$

in particular

$$\lim_A \int_n^\infty \mu_s * \sigma_\alpha ds = \int_n^\infty \mu_s * \sigma ds \quad \text{for } n \in \mathbb{N}.$$

Using the notation

$$\kappa_n = \int_0^n \mu_s ds \quad \text{for } n \in \mathbb{N},$$

we then have

$$\lim_A \kappa_n * \sigma_\alpha = \kappa_n * \sigma \quad \text{for all } n \in \mathbb{N}. \tag{3}$$

The inequality

$$\kappa * \sigma_\alpha \leqq \kappa * \tau \quad \text{for } \alpha \in A$$

shows that the net $(\sigma_\alpha)_{\alpha \in A}$ is vaguely bounded and hence vaguely relatively compact, cf. 15.7.

Let $(\sigma_\beta)_{\beta \in B}$ be an arbitrary vaguely convergent subnet of $(\sigma_\alpha)_{\alpha \in A}$ with limit σ'. Since the convolution of κ_n and the excessive measure $\kappa * \tau$ exists, cf. the proof of Theorem 16.7, we get by Proposition 15.5 that

$$\lim_B \kappa_n * \sigma_\beta = \kappa_n * \sigma' \quad \text{for all } n \in \mathbb{N},$$

so by (3) it follows that

$$\kappa_n * \sigma = \kappa_n * \sigma' \quad \text{for } n \in \mathbb{N}.$$

Letting $n \to \infty$ we find

$$\kappa * \sigma = \kappa * \sigma',$$

and by the principle of unicity of mass we finally get $\sigma = \sigma'$. Since every vaguely convergent subnet of the net $(\sigma_\alpha)_{\alpha \in A}$ has the same limit σ, it follows that

$$\lim_A \sigma_\alpha = \sigma \text{ vaguely.} \quad \square$$

16.11. Corollary. *For a transient convolution semigroup $(\mu_t)_{t>0}$ we have $\lim_{t\to\infty} \mu_t = 0$ vaguely.*

Proof. The net $(\kappa * \mu_t)_{t>0}$ converges vaguely to 0 as $t \to \infty$ and $\kappa * \mu_t \leqq \kappa$ for all $t>0$, so by Theorem 16.10 $\lim_{t\to\infty} \mu_t = 0$ vaguely. $\square$

16.12. Exercise. A convolution semigroup is transient if and only if there exists a non-invariant excessive measure.

This result can be used for another proof of the fact, cf. 14.1: A convolution semigroup $(\mu_t)_{t>0}$ on $\mathbb{R}$ supported by $[0, \infty[$ is either the degenerate semigroup $(\varepsilon_0)_{t>0}$ or it is transient.

16.13. Exercise. Let $\xi=\kappa*\sigma+\eta$ be the Riesz decomposition of the excessive measure ξ. Then σ and η are determined as the following vague limits:

$$\sigma=\lim_{\lambda\to\infty}\lambda(\xi-\lambda\rho_\lambda*\xi)$$

and

$$\eta=\lim_{\lambda\to 0}\lambda\rho_\lambda*\xi.$$

16.14. Theorem. *Let κ be the potential kernel for a transient convolution semigroup. For every open set $\omega\subseteq G$ and every $\sigma\in D^+(\kappa)$ there exists a κ-balayaged measure of σ on ω. In particular κ satisfies the balayage principle for all open sets.*

Proof. The convolution kernel $\lambda\kappa+\varepsilon_0$ is for $\lambda>0$ an elementary kernel (13.7), and it satisfies therefore in particular the balayage principle, cf. 15.23. The proportional kernel $\kappa+\frac{1}{\lambda}\varepsilon_0$ satisfies also this principle, and since $\kappa=\lim_{\lambda\to\infty}\left(\kappa+\frac{1}{\lambda}\varepsilon_0\right)$ vaguely, it follows from Proposition 15.14 that κ satisfies the balayage principle.

Let ω be an open, relatively compact subset of G and let $\sigma\in D^+(\kappa)$. In order to find a κ-balayaged measure of σ on ω we consider the monotonically increasing net $(\sigma_K)_{K\in\check{K}}$, where σ_K is the restriction of σ to the compact set K and $\check{K}$ is the directed set of compact subsets of G ordered by inclusion. For each $K\in\check{K}$ there exists a κ-balayaged measure σ'_K of σ_K on ω, and we have for all $K\in\check{K}$,

$$\operatorname{supp}(\sigma'_K)\subseteq\bar{\omega},$$
$$\kappa*\sigma'_K\leqq\kappa*\sigma_K\leqq\kappa*\sigma,$$
$$\kappa*\sigma'_K=\kappa*\sigma_K \text{ in } \omega,$$

which shows that the net $(\kappa*\sigma'_K)_{K\in\check{K}}$ is vaguely bounded. It follows by Theorem 16.10 that a vague accumulation point ξ for this net has the form $\xi=\kappa*\sigma'$, where σ' is a vague accumulation point for the net $(\sigma'_K)_{K\in\check{K}}$, and since

$$\lim_{\check{K}}\kappa*\sigma_K=\kappa*\sigma \text{ vaguely,}$$

(cf. 15.3) it follows that σ' is a κ-balayaged measure of σ on ω.

For $\sigma\in D^+(\kappa)$ and an arbitrary open set $\omega_0\subseteq G$ we consider the collection Ω of open, relatively compact sets ω such that $\omega\subseteq\omega_0$. There exists, by the first part of the proof, for every $\omega\in\Omega$ a κ-balayaged measure σ'_ω of σ on ω, i.e. a positive measure satisfying

$$\begin{aligned}&\operatorname{supp}(\sigma'_\omega)\subseteq\overline{\omega}\\&\kappa*\sigma'_\omega\leqq\kappa*\sigma\\&\kappa*\sigma'_\omega=\kappa*\sigma \text{ in } \omega.\end{aligned}$$

It follows that the family $(\kappa*\sigma'_\omega)_{\omega\in\Omega}$ is vaguely bounded. Let ξ' be a vague accumulation point for $(\kappa*\sigma'_\omega)_{\omega\in\Omega}$ as ω "tends to" ω_0. It follows from Theorem 16.10 that ξ' is of the form $\xi'=\kappa*\sigma'$ where $\sigma'\in D^+(\kappa)$ is supported by $\overline{\omega}_0$.

It is easily seen that σ' is a κ-balayaged measure of σ on ω_0. □

16.15. Definition. Let $\omega\subseteq G$ be an open set and let ξ be an excessive measure. The *reduced measure of* ξ *over* ω is the measure defined by

$$R_\xi^\omega=\inf\{\tau\mid\tau \text{ excessive}, \tau\geqq\xi \text{ on } \omega\}.$$

It is clear that R_ξ^ω is an excessive measure $\leqq\xi$, and $R_\xi^\omega=\xi$ on ω.

The mapping $(\omega,\xi)\overset{R}{\mapsto}R_\xi^\omega$ is increasing in both variables, i.e.

$$\omega_1\subseteq\omega_2 \;\Rightarrow\; R_\xi^{\omega_1}\leqq R_\xi^{\omega_2}$$

and

$$\xi_1\leqq\xi_2 \;\Rightarrow\; R_{\xi_1}^\omega\leqq R_{\xi_2}^\omega.$$

Moreover R has the following "continuity" properties:

16.16. Proposition. (i) *For a monotonically increasing net* $(\omega_\alpha)_{\alpha\in A}$ *of open subsets of* G *with* $\omega=\bigcup_{\alpha\in A}\omega_\alpha$ *and an excessive measure* ξ *we have*

$$\lim_A R_\xi^{\omega_\alpha}=\sup_A R_\xi^{\omega_\alpha}=R_\xi^\omega \text{ vaguely.}$$

(ii) *For a monotonically increasing net* $(\xi_\alpha)_{\alpha\in A}$ *of excessive measures with limit* $\xi=\sup_A\xi_\alpha$ *and an open subset* ω *of* G *we have*

$$\lim_A R_{\xi_\alpha}^\omega=\sup_A R_{\xi_\alpha}^\omega=R_\xi^\omega \text{ vaguely.}$$

Proof. (i) The net $(R_\xi^{\omega_\alpha})_{\alpha\in A}$ is monotonically increasing and $R_\xi^{\omega_\alpha}\leqq R_\xi^\omega$ for $\alpha\in A$, and therefore

$$\tau=\lim_A R_\xi^{\omega_\alpha}=\sup_A R_\xi^{\omega_\alpha}$$

exists and τ is an excessive measure $\leqq R_\xi^\omega$. In order to establish that $\tau\geqq R_\xi^\omega$, it suffices to prove that $\tau=\xi$ on ω. Suppose that $f\in C_c(G)$ has its support contained in ω. Then there exists $\alpha_0\in A$ such that

$$\operatorname{supp}(f)\subseteq\omega_\alpha \quad \text{for } \alpha\geqq\alpha_0,$$

and it follows that

$$\langle R_\xi^{\omega_\alpha}, f\rangle = \langle \xi, f\rangle \quad \text{for } \alpha \geqq \alpha_0,$$

which shows that $\langle \tau, f\rangle = \langle \xi, f\rangle$.

(ii) The net $(R_{\xi_\alpha}^\omega)_{\alpha\in A}$ is monotonically increasing and $R_{\xi_\alpha}^\omega \leqq R_\xi^\omega$ for $\alpha\in A$, and therefore

$$\tau' = \lim_A R_{\xi_\alpha}^\omega = \sup_A R_{\xi_\alpha}^\omega$$

exists and τ' is an excessive measure $\leqq R_\xi^\omega$. It is easily seen that $\tau' = \xi$ on ω, and therefore $\tau' \geqq R_\xi^\omega$. □

16.17. Proposition. *Let ξ be an excessive measure and $\omega \subseteq G$ an open set. For the Riesz decomposition of R_ξ^ω,*

$$R_\xi^\omega = \kappa * \sigma + \eta,$$

we have $\operatorname{supp}(\sigma) \subseteq \bar{\omega}$.

Proof. Let σ' be a κ-balayaged measure of σ on ω; such a measure exists by Theorem 16.14. Then

$$\tau = \kappa * \sigma' + \eta$$

is an excessive measure satisfying

$$\tau \leqq R_\xi^\omega$$

and

$$\tau = R_\xi^\omega = \xi \quad \text{on } \omega.$$

This shows that $\tau \geqq R_\xi^\omega$ hence $\tau = R_\xi^\omega$. By the uniqueness of the Riesz decomposition (16.7) we then have $\sigma = \sigma'$, in particular $\operatorname{supp}(\sigma) \subseteq \bar{\omega}$. □

16.18. Corollary. *Let $\omega \subseteq G$ be an open set and let $\sigma \in D^+(\kappa)$. The reduced measure of $\kappa * \sigma$ over ω is a potential generated by a κ-balayaged measure of σ on ω, denoted σ^ω, i.e.*

$$R_{\kappa*\sigma}^\omega = \kappa * \sigma^\omega.$$

Proof. Since $R_{\kappa*\sigma}^\omega$ is dominated by the potential $\kappa * \sigma$ there exists by 16.9 a (uniquely determined) measure $\sigma^\omega \in D^+(\kappa)$ such that

$$R_{\kappa*\sigma}^\omega = \kappa * \sigma^\omega,$$

and from 16.17 and the properties of the reduced measure it is obvious that σ^ω is a κ-balayaged measure of σ on ω. □

16.19. Definition. For $\sigma \in D^+(\kappa)$ and an open set $\omega \subseteq G$, the uniquely determined measure $\sigma^\omega \in D^+(\kappa)$ such that

$$R_{\kappa*\sigma}^\omega = \kappa * \sigma^\omega$$

is called the *canonical κ-balayaged measure of σ on ω.*

16.20. Proposition. (i) *For a monotonically increasing net $(\omega_\alpha)_{\alpha\in A}$ of open subsets of G with $\omega=\bigcup_{\alpha\in A}\omega_\alpha$ and for $\sigma\in D^+(\kappa)$ we have*

$$\lim_A \sigma^{\omega_\alpha}=\sigma^{\omega} \quad \textit{vaguely.}$$

(ii) *Let $\sigma\in D^+(\kappa)$ and let $(\sigma_\alpha)_{\alpha\in A}$ be a monotonically increasing net such that $\sigma=\lim_A \sigma_\alpha$ vaguely. For every open set $\omega\subseteq G$ we have*

$$\lim_A \sigma_\alpha^{\omega}=\sigma^{\omega} \quad \textit{vaguely.}$$

Proof. The assertions follow easily from 16.16, 16.10 and 15.3. ▯

16.21. Proposition. *The reduced measure R_ξ^ω of an excessive measure ξ over an open, relatively compact set $\omega\subseteq G$ is a potential.*

Proof. If the measure ξ has a continuous density f with respect to the Haar measure ω_G on G, $\xi=f\omega_G$, there exists a function $\varphi\in C_c^+(G)$ such that

$$\kappa*\varphi\geqq f \quad \text{on } \bar{\omega},$$

(cf. 1.9), and the excessive measure R_ξ^ω is thus dominated by the potential $\kappa*(\varphi\,\omega_G)$ and hence a potential by 16.9.

Let V_0 be a compact neighbourhood of zero in G and let $(\varphi_V)_{V\in\hat{V}}$ be an approximate unit, where $\hat{V}$ consists of the compact neighbourhoods of 0 which are contained in V_0. For $V\in\hat{V}$ the excessive measure $\xi*\varphi_V$ has a continuous density with respect to ω_G, and by the first part of the proof and Proposition 16.17 there exists a positive measure τ_V such that $\operatorname{supp}(\tau_V)\subseteq\bar{\omega}$ and

$$R^\omega_{\xi*\varphi_V}=\kappa*\tau_V. \tag{4}$$

The net $(\kappa*\tau_V)_{V\in\hat{V}}$ is vaguely bounded, because for $g\in C_c^+(G)$ and $V\in\hat{V}$ we find

$$\langle\kappa*\tau_V, g\rangle\leqq\langle\xi*\varphi_V, g\rangle=\langle\xi*\check{g}, \check{\varphi}_V\rangle\leqq\sup_{x\in -V_0}\xi*\check{g}(x)<\infty.$$

By Lemma 15.7 the net $(\tau_V)_{V\in\hat{V}}$ is also vaguely bounded, and there exists consequently a subnet $(\tau_{V_\alpha})_{\alpha\in A}$ which converges vaguely to a measure τ'.

Since the measures τ_{V_α} are supported by the fixed compact set $\bar{\omega}$ we have

$$\lim_A \kappa*\tau_{V_\alpha}=\kappa*\tau' \quad \text{vaguely},$$

and we clearly have

$$\lim_A \xi*\varphi_{V_\alpha}=\xi \quad \text{vaguely},$$

so by (4)

$$\kappa*\tau'=\xi \quad \text{on } \omega.$$

It follows that

$$R_\xi^\omega\leqq\kappa*\tau',$$

and the measure R_ξ^ω is by 16.9 a potential. ▯

An application of Proposition 16.21 with $\xi=\omega_G$, the Haar measure of G, gives that for every open relatively compact subset $\omega\subseteq G$ the reduced measure $R^{\omega}_{\omega_G}$ is a potential, i.e. there exists a measure $\lambda_\omega\in D^+(\kappa)$ such that

$$R^{\omega}_{\omega_G}=\kappa*\lambda_\omega. \tag{5}$$

This proves the following theorem.

16.22. Theorem. *The potential kernel satisfies the* equilibrium principle, *i.e. for every open relatively compact set $\omega\subseteq G$ there exists a measure $\lambda'_\omega\in D^+(\kappa)$ such that*

(i) $\operatorname{supp}(\lambda'_\omega)\subseteq\bar{\omega}$.

(ii) $\kappa*\lambda'_\omega\leqq\omega_G$.

(iii) $\kappa*\lambda'_\omega=\omega_G$ in ω.

Any measure λ'_ω with these properties is called a κ-equilibrium distribution *for ω.*

16.23. Definition. The uniquely determined measure $\lambda_\omega\in D^+(\kappa)$ such that

$$R^{\omega}_{\omega_G}=\kappa*\lambda_\omega$$

is called the *canonical κ-equilibrium distribution* for the open, relatively compact set $\omega\subseteq G$. The number

$$\operatorname{cap}(\omega)=\int d\lambda_\omega$$

is called the *κ-capacity* of ω.

16.24. Corollary. *Let $\omega\subseteq G$ be an open, relatively compact subset of G. For every compact neighbourhood V of 0 in G there exists $\varphi\in C_c^+(G)$ such that*

(i) $\operatorname{supp}(\varphi)\subseteq\bar{\omega}+V$,

(ii) $\kappa*\varphi\leqq1$,

(iii) $\kappa*\varphi=1$ in ω.

Proof. For compact neighbourhoods V and W of 0 in G such that $W-W\subseteq V$ we choose a κ-equilibrium distribution σ for $\omega-W$ and $f\in C_c^+(G)$ such that

$$\operatorname{supp}(f)\subseteq W \quad\text{and}\quad \langle\omega_G,f\rangle=1.$$

The function $\varphi=\sigma*f$ clearly satisfies (i), (ii) and (iii) of the corollary. □

16.25. Theorem. *The potential kernel κ satisfies the* principle of positivity of mass, *i.e. for all $\sigma_1,\sigma_2\in D^+(\kappa)$ we have the implication:*

$$\kappa*\sigma_1\leqq\kappa*\sigma_2 \;\Rightarrow\; \sigma_1(G)\leqq\sigma_2(G).$$

Proof. The reflected measure $\check{\kappa}$ is potential kernel for the transient convolution semigroup $(\check{\mu}_t)_{t>0}$. Let ω be an open, relatively compact subset of G. By Corollary 16.24 there exists a function $\varphi\in C_c^+(G)$ such that $\check{\kappa}*\varphi\leqq1$ and $\check{\kappa}*\varphi=1$ in

ω. If $\sigma_1, \sigma_2 \in D^+(\kappa)$ satisfies $\kappa * \sigma_1 \leqq \kappa * \sigma_2$, then we get

$$\sigma_1(\omega) \leqq \langle \sigma_1, \check{\kappa} * \varphi \rangle = \langle \kappa * \sigma_1, \varphi \rangle \leqq \langle \kappa * \sigma_2, \varphi \rangle = \langle \sigma_2, \check{\kappa} * \varphi \rangle \leqq \sigma_2(G),$$

and since ω is arbitrary, it follows that

$$\sigma_1(G) \leqq \sigma_2(G). \quad \square$$

16.26. Corollary. *The* κ*-balayage* diminishes the total mass, *i.e. for every* $\sigma \in D^+(\kappa)$ *and every* κ*-balayaged measure* σ' *of* σ *on an open set* $\omega \subseteq G$ *we have*

$$\sigma'(G) \leqq \sigma(G).$$

16.27. Corollary. *The potential kernel* κ *satisfies the complete maximum principle.*

Proof. Let $f, g \in C_c^+(G)$ and $a \geqq 0$ and suppose (cf. 15.10) that

$$\kappa * f(x) \leqq \kappa * g(x) + a \qquad \text{for all } x \in \operatorname{supp}(f).$$

In order to see that this inequality holds for an arbitrary $x \in G$, we put $\omega = \{f > 0\}$ and consider the canonical $\check{\kappa}$-balayaged measure ε_x^ω of ε_x on ω. This measure exists because $\check{\kappa}$ is potential kernel for the convolution semigroup $(\check{\mu}_t)_{t>0}$, cf. 16.19, and it satisfies $\varepsilon_x^\omega(G) \leqq 1$, cf. 16.26. We then have

$$\begin{aligned}\kappa * f(x) &= \langle \varepsilon_x, \kappa * f \rangle = \langle \check{\kappa} * \varepsilon_x, f \rangle = \langle \check{\kappa} * \varepsilon_x^\omega, f \rangle \\ &= \langle \varepsilon_x^\omega, \kappa * f \rangle \leqq \langle \varepsilon_x^\omega, \kappa * g \rangle + a = \langle \check{\kappa} * \varepsilon_x^\omega, g \rangle + a \\ &\leqq \langle \check{\kappa} * \varepsilon_x, g \rangle + a = \kappa * g(x) + a. \quad \square\end{aligned}$$

16.28. Remark. Let $f \in C_c^+(G)$. Putting

$$a = \sup\{\kappa * f(x) \mid x \in \operatorname{supp}(f)\}$$

we clearly have

$$\kappa * f(x) \leqq a \qquad \text{for } x \in \operatorname{supp}(f),$$

so by 16.27 we conclude that

$$\kappa * f(x) \leqq a \qquad \text{for all } x \in G.$$

This gives another proof of the result, that κ is shift-bounded, cf. 13.10.

16.29. Proposition. *Every excessive measure* ξ *is the vague limit of a monotonically increasing net of potentials.*

Proof. Let A be the collection of all open, relatively compact subsets of G ordered by inclusion. It follows from Proposition 16.21 that $(R_\xi^\omega)_{\omega \in A}$ is a monotonically increasing net of potentials, and by Proposition 16.16 (i) this net tends to $R_\xi^G = \xi$ vaguely. $\square$

It follows from 16.14 (or 16.27) that the kernel κ satisfies the domination principle, cf. 15.13, and this principle can be extended in the following way:

16.30. Proposition. *Let $\sigma \in D^+(\kappa)$ and let ξ be an excessive measure. If there exists an open set $\omega \subseteq G$ such that*

(i) $\operatorname{supp}(\sigma) \subseteq \omega$
(ii) $\kappa * \sigma \leqq \xi$ on ω,

then $\kappa * \sigma \leqq \xi$.

Proof. Suppose first that σ has compact support, and it is then no restriction to suppose also that $\bar{\omega}$ is compact.

Putting

$$\xi_0 = \inf(\kappa * \sigma, \xi),$$

the reduced measure $R^{\omega}_{\xi_0}$ is dominated by a potential and hence a potential of the form

$$R^{\omega}_{\xi_0} = \kappa * \tau,$$

where $\operatorname{supp}(\tau) \subseteq \bar{\omega}$, cf. 16.17, and we have

$$\kappa * \sigma = \xi_0 = \kappa * \tau \quad \text{on } \omega.$$

By hypothesis $\operatorname{supp}(\sigma)$ is a compact subset of ω and there exists consequently a compact symmetric neighbourhood V_0 of 0 in G such that

$$\operatorname{supp}(\sigma) + V_0 + V_0 \subseteq \omega.$$

For an approximate unit $(\varphi_V)_{V \in \grave{V}}$ we have for $V \subseteq V_0$ that

$$\kappa * (\sigma * \varphi_V)(x) = \kappa * (\tau * \varphi_V)(x) \quad \text{for } x \in \operatorname{supp}(\sigma * \varphi_V)$$

because

$$\operatorname{supp}(\varepsilon_x * \check{\varphi}_V) \subseteq x - V_0 \subseteq \operatorname{supp}(\sigma) + V_0 - V_0 \subseteq \omega$$

for $x \in \operatorname{supp}(\sigma * \varphi_V)$.

By the domination principle for κ it follows that

$$\kappa * (\sigma * \varphi_V) \leqq \kappa * (\tau * \varphi_V)$$

for $V \in \grave{V}$ such that $V \subseteq V_0$, and taking limits along $\grave{V}$ we get

$$\kappa * \sigma \leqq \kappa * \tau$$

and hence

$$\kappa * \sigma \leqq R^{\omega}_{\xi_0} \leqq \xi_0 \leqq \xi.$$

Suppose now that $\sigma \in D^+(\kappa)$ is arbitrary and satisfies (i) and (ii). Denoting by $\grave{K}$ the upward filtering (by inclusion) set of compact subsets of G, the restriction

σ_K of σ to $K \in \check{K}$ satisfies

$$\operatorname{supp}(\sigma_K) \subseteq \omega \quad \text{and} \quad \kappa * \sigma_K \leqq \xi \quad \text{on } \omega,$$

and from what is already proved we get

$$\kappa * \sigma_K \leqq \xi \quad \text{on } G \text{ for all } K \in \check{K}.$$

By Lemma 15.3 we have

$$\kappa * \sigma = \lim_{\check{K}} \kappa * \sigma_K \quad \text{vaguely}$$

and hence

$$\kappa * \sigma \leqq \xi. \quad \square$$

16.31. Proposition. *For an open set $\omega \subseteq G$ and excessive measures ξ_1 and ξ_2 we have*

$$R^{\omega}_{\xi_1+\xi_2} = R^{\omega}_{\xi_1} + R^{\omega}_{\xi_2},$$

and for $\sigma_1, \sigma_2 \in D^+(\kappa)$ we have

$$(\sigma_1 + \sigma_2)^{\omega} = \sigma_1^{\omega} + \sigma_2^{\omega}.$$

Proof. The measure $R^{\omega}_{\xi_1} + R^{\omega}_{\xi_2}$ is an excessive measure which is equal to $\xi_1 + \xi_2$ in ω, and we therefore have

$$R^{\omega}_{\xi_1+\xi_2} \leqq R^{\omega}_{\xi_1} + R^{\omega}_{\xi_2}.$$

Let A be the directed set of open relatively compact sets $\alpha \subseteq G$ such that $\bar{\alpha} \subseteq \omega$, ordered by inclusion. For $\alpha, \beta \in A$ the measure

$$\zeta = R^{\alpha}_{\xi_1} + R^{\beta}_{\xi_2}$$

is a potential $\zeta = \kappa * \sigma$ where $\operatorname{supp}(\sigma) \subseteq \bar{\alpha} \cup \bar{\beta} \subseteq \omega$. Since $\zeta \leqq \xi_1 + \xi_2$, it follows that

$$\zeta = \kappa * \sigma \leqq R^{\omega}_{\xi_1+\xi_2} \quad \text{on } \omega,$$

and by Proposition 16.30 we then get

$$\zeta = R^{\alpha}_{\xi_1} + R^{\beta}_{\xi_2} \leqq R^{\omega}_{\xi_1+\xi_2}.$$

If in this inequality we let α "increase" to ω for fixed $\beta \in A$ and then let β "increase" to ω, it follows by Proposition 16.16 (i) that

$$R^{\omega}_{\xi_1} + R^{\omega}_{\xi_2} \leqq R^{\omega}_{\xi_1+\xi_2}.$$

The second part of the Proposition is an immediate consequence of the first part and the principle of unicity of mass for κ, cf. 16.8 and 16.19. $\square$

16.32. Exercise. Let ξ be an excessive measure and let $\omega \subseteq \Omega \subseteq G$ be open sets. Then

$$R^{\omega}_{R^{\Omega}_{\xi}} = R^{\Omega}_{R^{\omega}_{\xi}} = R^{\omega}_{\xi},$$

and for $\sigma \in D^+(\kappa)$ we have

$$(\sigma^{\omega})^{\Omega} = (\sigma^{\Omega})^{\omega} = \sigma^{\omega}.$$

16.33. Exercise. The capacity $\omega \mapsto \operatorname{cap}(\omega)$, defined for all open, relatively compact subsets of G, has the following properties:

(i) $\operatorname{cap}(\emptyset) = 0$ and $\operatorname{cap}(\omega) > 0$ if $\omega \neq \emptyset$.

(ii) $\omega_1 \subseteq \omega_2 \Rightarrow \operatorname{cap}(\omega_1) \leqq \operatorname{cap}(\omega_2)$.

(iii) For a monotonically increasing net $(\omega_\alpha)_{\alpha \in A}$ of open, relatively compact subsets of G such that $\omega = \bigcup_{\alpha \in A} \omega_\alpha$ is relatively compact we have

$$\operatorname{cap}(\omega) = \lim_A \operatorname{cap}(\omega_\alpha) = \sup_A \operatorname{cap}(\omega_\alpha).$$

(iv) $\operatorname{cap}(\omega_1 \cup \omega_2) + \operatorname{cap}(\omega_1 \cap \omega_2) \leqq \operatorname{cap}(\omega_1) + \operatorname{cap}(\omega_2)$,

(i.e. the capacity is *strongly subadditive*).

16.34. Most of the material in this paragraph is taken from the papers of Deny [1] and [5], but many of the proofs are different from the proofs given by Deny.

§ 17. Fundamental Families Associated With Potential Kernels

In this paragraph we introduce the notion of a fundamental family associated with a convolution kernel on the LCA-group G and prove that the convolution kernels for which there exists an associated fundamental family—they are called perfect kernels—are precisely the potential kernels for transient convolution semigroups. Then it is shown how excessive and invariant measures with respect to a transient convolution semigroup can be characterized by means of a fundamental family associated with the potential kernel for the convolution semigroup, and finally some examples of fundamental families are given.

17.1. Definition. A *fundamental family associated with* a positive measure κ on G is a net $(\sigma_V)_{V \in \check{V}}$, indexed by a base $\check{V}$ of compact neighbourhoods of 0, consisting of positive measures on G verifying for all $V \in \check{V}$,

(i) $\sigma_V \in D^+(\kappa)$, $\sigma_V * \kappa \leqq \kappa$ and $\sigma_V * \kappa \neq \kappa$,

(ii) $\sigma_V * \kappa = \kappa$ on $\complement V$,

(iii) $\lim_{n \to \infty} \sigma_V^n * \kappa = 0$ vaguely,

(iv) $\sigma_V(G) \leqq 1$.

17.2. Definition. A positive measure κ on G is called a *perfect kernel*, if there exists a fundamental family associated with κ.

It follows from (i) that a perfect kernel is different from 0.

In general there exist different fundamental families associated with a given perfect kernel κ. If $(\sigma_V)_{V\in\check{\mathfrak{V}}}$ is a fundamental family associated with κ and if $\check{\mathfrak{W}}\subseteq\check{\mathfrak{V}}$ is a base of compact neighbourhoods of 0, then $(\sigma_V)_{V\in\check{\mathfrak{W}}}$ is a fundamental family associated with κ.

On the other hand, given a perfect kernel κ, there always exists a fundamental family associated with κ, for which the index set consists of all compact neighbourhoods of 0. This follows from 17.5 and 17.8 below.

17.3. Example. Every elementary kernel κ is a perfect kernel. In fact, there exists a positive measure μ on G with $\mu(G)\leqq 1$ such that

$$\kappa=\sum_{n=0}^{\infty}\mu^n,$$

and the constant net $(\mu)_{K\in\check{\mathfrak{K}}(0)}$, where $\check{\mathfrak{K}}(0)$ is the set of all compact neighbourhoods of 0, is easily seen to be a fundamental family associated with κ.

More generally we have the following result.

17.4. Theorem. *The potential kernel κ for a transient convolution semigroup $(\mu_t)_{t>0}$ on G is a perfect kernel.*

Proof. Let $\check{\mathfrak{V}}$ be a base of compact neighbourhoods of 0 and choose for $V\in\check{\mathfrak{V}}$ a κ-balayaged measure σ_V of ε_0 on $\complement V$, cf. 16.14. We shall see that $(\sigma_V)_{V\in\check{\mathfrak{V}}}$ is a fundamental family associated with κ. For $V\in\check{\mathfrak{V}}$ we have

$$\sigma_V*\kappa\leqq\kappa \quad\text{and}\quad \sigma_V*\kappa=\kappa \text{ on } \complement V.$$

Since $\operatorname{supp}(\sigma_V)\subseteq\overline{\complement V}$ it is clear that $\varepsilon_0\neq\sigma_V$, and as κ satisfies the principle of unicity of mass (cf. 16.8) it follows that (i) and (ii) of Definition 17.1 are satisfied. The condition (iv) is satisfied because the κ-balayage diminishes the total mass, cf. 16.26.

The sequence $(\sigma_V^n*\kappa)_{n\in\mathbb{N}}$ is decreasing by (i), and it follows that the vague limit

$$\xi=\lim_{n\to\infty}\sigma_V^n*\kappa$$

exists, and ξ is a potential, i.e. $\xi=\kappa*\sigma$ for a $\sigma\in D^+(\kappa)$. By Lemma 15.3 we find

$$\xi=\sigma_V^p*\xi \qquad \text{for } p\in\mathbb{N},$$

and since

$$\xi\leqq\sigma_V^n*\kappa\leqq\kappa \qquad \text{for } n\in\mathbb{N},$$

the convolution $\xi*\sigma$ exists and

$$\xi*\sigma=\lim_{n\to\infty}\sigma_V^n*\kappa*\sigma=\lim_{n\to\infty}\sigma_V^n*\xi=\xi.$$

We consequently have

$$(\kappa-\xi)*\sigma=\kappa*\sigma-\xi*\sigma=\xi-\xi=0, \tag{1}$$

and since

$$\kappa-\xi \geqq \kappa-\sigma_V*\kappa$$

it follows that $\kappa-\xi$ is positive and non-zero, so by (1) we get that $\sigma=0$ and hence $\xi=0$, which shows that condition (iii) of Definition 17.1 is satisfied. □

17.5. Remark. From the proof of Theorem 17.4 we see that a fundamental family $(\sigma_V)_{V\in\check{V}}$ associated with the potential kernel κ (where $\check{V}$ is an arbitrary base of neighbourhoods of 0 consisting of compact sets) can be obtained by choosing for every $V\in\check{V}$ a κ-balayaged measure σ_V of ε_0 on the open set $\complement V$.

In particular $(\varepsilon_0^{\complement V})_{V\in\check{K}(0)}$ is a fundamental family associated with κ, where $\check{K}(0)$ is the set of all compact neighbourhoods of 0.

17.6. Let κ be a perfect kernel with an associated fundamental family $(\sigma_V)_{V\in\check{V}}$. For every $V\in\check{V}$ the measure

$$\kappa-\sigma_V*\kappa$$

is a non-zero positive measure with compact support contained in V and there exists consequently a uniquely determined positive number $a_V>0$ such that the measure

$$\eta_V=a_V(\kappa-\sigma_V*\kappa)=a_V\kappa*(\varepsilon_0-\sigma_V) \tag{2}$$

has total mass 1.

Let $V\in\check{V}$. For $N\in\mathbb{N}$ we have

$$\left(\sum_{n=0}^{N}\sigma_V^n\right)*(\kappa-\sigma_V*\kappa)=\kappa-\sigma_V^{N+1}*\kappa,$$

and hence

$$\left(\frac{1}{a_V}\sum_{n=0}^{N}\sigma_V^n\right)*\eta_V=\kappa-\sigma_V^{N+1}*\kappa\leqq\kappa, \tag{3}$$

so the sequence $\left(\frac{1}{a_V}\sum_{n=0}^{N}\sigma_V^n\right)_{N\in\mathbb{N}}$ is by Lemma 15.7 vaguely bounded and therefore vaguely convergent. We define

$$\kappa_V=\frac{1}{a_V}\sum_{n=0}^{\infty}\sigma_V^n \quad \text{for } V\in\check{V}, \tag{4}$$

and letting $N\to\infty$ in (3) we get by 17.1(iii)

$$\kappa_V*\eta_V=\kappa \quad \text{for } V\in\check{V}. \tag{5}$$

17.7. Lemma. *A perfect kernel κ is shift-bounded.*

Proof. For $V \in \check{V}$, the measure $a_V \kappa_V$ is an elementary kernel and in particular a potential kernel for a transient convolution semigroup, cf. 13.6, so by Proposition 13.10, $a_V \kappa_V$ is shift-bounded.

For $f \in C_c^+(G)$ we have $\eta_V * f \in C_c^+(G)$, because η_V has compact support, and then

$$\kappa * f = \kappa_V * (\eta_V * f)$$

is bounded. □

The following result is a special case of a theorem of Deny [5]. The dual group of G is (as usual) denoted Γ.

17.8. Theorem. *Every perfect kernel κ is the potential kernel for a uniquely determined transient convolution semigroup.*

Proof. We shall start with the construction of a continuous negative definite function ψ on Γ and then show that the convolution semigroup $(\mu_t)_{t>0}$ on G associated with ψ is transient and has κ as potential kernel.

Let $(\sigma_V)_{V \in \check{V}}$ be a fundamental family associated with κ. Using the notation from 17.6 we have that

$$\lim_{\check{V}} \eta_V = \varepsilon_0$$

in the Bernoulli topology, so by Theorem 3.13

$$\lim_{\check{V}} \hat{\eta}_V = 1 \tag{6}$$

uniformly on compact subsets of Γ.

For $V, V' \in \check{V}$ we have

$$a_{V'} \eta_V * (\varepsilon_0 - \sigma_{V'}) = a_V \eta_{V'} * (\varepsilon_0 - \sigma_V),$$

which by Fourier transformation gives

$$a_{V'} \hat{\eta}_V (1 - \hat{\sigma}_{V'}) = a_V \hat{\eta}_{V'} (1 - \hat{\sigma}_V). \tag{7}$$

There exists a uniquely determined function $\psi \colon \Gamma \to \mathbb{C}$ such that

$$\hat{\eta}_V(\gamma) \psi(\gamma) = a_V (1 - \hat{\sigma}_V(\gamma)) \quad \text{for } \gamma \in \Gamma \text{ and } V \in \check{V}. \tag{8}$$

To see this let $\gamma \in \Gamma$ and choose $V \in \check{V}$ such that $\hat{\eta}_V(\gamma) \neq 0$ (this is possible by (6)). The number $\psi(\gamma) \in \mathbb{C}$ given by

$$\psi(\gamma) = \frac{a_V (1 - \hat{\sigma}_V(\gamma))}{\hat{\eta}_V(\gamma)} \tag{9}$$

is independent of $V \in \check{V}$ such that $\hat{\eta}_V(\gamma) \neq 0$, cf. (7). With this definition of $\psi(\gamma)$, (8) holds for all $V \in \check{V}$ such that $\hat{\eta}_V(\gamma) \neq 0$, but if $\hat{\eta}_V(\gamma) = 0$ we may choose $V' \in \check{V}$ such that $\hat{\eta}_{V'}(\gamma) \neq 0$, and it follows by (7) that $\hat{\sigma}_V(\gamma) = 1$, so that (8) holds also in this case.

The function ψ is clearly continuous, and from Corollary 7.7 it follows that $\hat{\eta}_V \psi$ is negative definite. By (6) we have

$$\lim_{\check{V}} \hat{\eta}_V \psi = \psi$$

uniformly over compact subsets of Γ, and ψ is therefore negative definite.

Let $(\mu_t)_{t>0}$ be the convolution semigroup on G associated with ψ and let $(\rho_\lambda)_{\lambda>0}$ be the resolvent for $(\mu_t)_{t>0}$. Then we have for $V \in \check{V}$ and $\lambda > 0$

$$\eta_V - a_V \rho_\lambda * (\varepsilon_0 - \sigma_V) = \lambda \rho_\lambda * \eta_V. \tag{10}$$

In fact, the Fourier transform of the left-hand side of (10) is equal to the Fourier transform of the right-hand side of (10):

$$\hat{\eta}_V - a_V \hat{\rho}_\lambda (1 - \hat{\sigma}_V) = \hat{\eta}_V - \frac{a_V (1 - \hat{\sigma}_V)}{\psi + \lambda} = \frac{\lambda \hat{\eta}_V}{\psi + \lambda} = (\lambda \rho_\lambda * \eta_V)\hat{}.$$

The convolution of κ and ρ_λ exists, because κ is shift-bounded and ρ_λ is bounded, cf. 1.13, and (10) gives by convolution with κ that

$$\kappa * \eta_V - \rho_\lambda * \eta_V = \lambda \rho_\lambda * \kappa * \eta_V \qquad \text{for } V \in \check{V} \text{ and } \lambda > 0.$$

Letting V "shrink" to $\{0\}$ through $\check{V}$ we get

$$\kappa - \rho_\lambda = \lambda \rho_\lambda * \kappa \qquad \text{for } \lambda > 0,$$

so in particular $\rho_\lambda \leqq \kappa$ for $\lambda > 0$, which implies that $(\mu_t)_{t>0}$ is transient (cf. 13.1) and that the potential kernel $\rho_0 = \lim_{\lambda \to 0} \rho_\lambda$ satisfies $\rho_0 \leqq \kappa$.

The proof will be finished, when we have seen that $\rho_0 = \kappa$. Letting $\lambda \downarrow 0$ in (10) we get by Lemma 15.3, since $\rho_\lambda * \sigma_V \leqq \kappa * \sigma_V$, that

$$\eta_V - a_V \rho_0 + a_V \rho_0 * \sigma_V = 0 \qquad \text{for } V \in \check{V},$$

or

$$(\kappa - \rho_0) * \sigma_V = \kappa - \rho_0.$$

By repeated convolution with σ_V we find for every $n \in \mathbb{N}$

$$0 \leqq \kappa - \rho_0 = (\kappa - \rho_0) * \sigma_V^n \leqq \kappa * \sigma_V^n,$$

where the right-hand side tends to zero vaguely as $n \to \infty$ by 17.1 (iii), and it follows that $\kappa = \rho_0$.

By 15.21 there exists at most one transient convolution semigroup on G with κ as potential kernel. □

The convolution semigroup $(\mu_t)_{t>0}$ from Theorem 17.8 can be obtained as a limit of convolution semigroups, which are determined by the measures in the fundamental family $(\sigma_V)_{V\in\check{\mathscr{V}}}$. Using the notation from 17.6 the precise form of this approximation can be formulated as follows:

17.9. Proposition. *Let κ be a perfect kernel and $(\sigma_V)_{V\in\check{\mathscr{V}}}$ a fundamental family associated with κ and let $(\mu_t)_{t>0}$ be the uniquely determined transient convolution semigroup on G with κ as potential kernel.*

Defining for $V\in\check{\mathscr{V}}$ the convolution semigroup $(\mu_t^V)_{t>0}$ by

$$\mu_t^V = e^{-ta_V}\exp(t a_V \sigma_V) \quad \textit{for } t>0, \tag{11}$$

we have for each $t>0$

$$\lim_{\check{\mathscr{V}}} \mu_t^V = \mu_t$$

in the Bernoulli topology.

Proof. It follows by 8.10 that $(\mu_t^V)_{t>0}$ is a convolution semigroup and the Fourier transform of μ_t^V is

$$\mathscr{F}(\mu_t^V) = e^{-ta_V(1-\hat{\sigma}_V)} = e^{-t\hat{\eta}_V\psi} \quad \text{for } t>0 \text{ and } V\in\check{\mathscr{V}},$$

which by (6) tends to $e^{-t\psi}$ uniformly over compact subsets of Γ, when V runs through $\check{\mathscr{V}}$. The assertion follows now by Theorem 3.13. □

The potential kernels for transient convolution semigroups or equivalently the perfect kernels, have been characterized by Choquet and Deny in another manner:

17.10. Theorem. *Every shift-bounded convolution kernel κ on G satisfying the balayage principle for all open sets and the principle of unicity of mass is the potential kernel for a transient convolution semigroup.*

For the *proof* we refer to Choquet and Deny [3].

For convolution kernels on $\mathbb{R}$ supported by $[0,\infty[$ a stronger result is available:

17.11. Theorem. *Let κ be a non-zero shift-bounded convolution kernel on $\mathbb{R}$ such that* $\operatorname{supp}(\kappa)\subseteq[0,\infty[$ *and suppose that κ verifies the balayage principle. Then κ is the potential kernel for a transient convolution semigroup on $\mathbb{R}$ supported by $[0,\infty[$.*

Proof. We first remark, that the convolution of any two positive measures supported by $[0,\infty[$ exists.

For $\varepsilon\in]0,1[$ and $n\in\mathbb{N}$ we denote by $\sigma_{\varepsilon,n}$ a κ-balayaged measure of ε_0 on the open interval $]\varepsilon, n[$.

Let $\varepsilon\in]0,1[$ be fixed. Since

$$\kappa * \sigma_{\varepsilon,n} \leqq \kappa \quad \text{for } n\in\mathbb{N},$$

Lemma 15.7 implies that the sequence $(\sigma_{\varepsilon,n})_{n\in\mathbb{N}}$ is vaguely bounded, and there exists consequently a measure σ_ε and a subsequence $(\sigma_{\varepsilon,n_p})_{p\in\mathbb{N}}$ such that

$$\lim_{p\to\infty} \sigma_{\varepsilon,n_p} = \sigma_\varepsilon \text{ vaguely.}$$

It is clear that $\operatorname{supp}(\sigma_\varepsilon) \subseteq [\varepsilon, \infty[$, and since the convolution $\kappa * \kappa$ exists, it follows by Proposition 15.5 that

$$\lim_{p\to\infty} \kappa * \sigma_{\varepsilon,n_p} = \kappa * \sigma_\varepsilon \text{ vaguely,}$$

and we easily get

$$\kappa * \sigma_\varepsilon \leqq \kappa \quad \text{and} \quad \kappa * \sigma_\varepsilon = \kappa \text{ in }]\varepsilon, \infty[, \tag{12}$$

which shows that σ_ε is a κ-balayaged measure of ε_0 on $]\varepsilon, \infty[$.

Let $\check{V}$ denote the base $\{[-\varepsilon, \varepsilon] \mid \varepsilon \in]0, 1[\}$ of compact neighbourhoods of 0 in $\mathbb{R}$. We shall now see that the net $(\sigma_\varepsilon)_{\varepsilon\in]0,1[}$ is a fundamental family associated with κ. (For convenience the index set $\check{V}$ has been replaced by the set $]0, 1[$, which is in one-to-one correspondence with $\check{V}$.)

If $\kappa * \sigma_\varepsilon = \kappa$ for an $\varepsilon \in]0, 1[$ then

$$\operatorname{supp}(\kappa) = \operatorname{supp}(\kappa * \sigma_\varepsilon) = \operatorname{supp}(\kappa) + \operatorname{supp}(\sigma_\varepsilon),$$

(note that $\operatorname{supp}(\kappa) + \operatorname{supp}(\sigma_\varepsilon)$ is closed as the sum of two closed subsets of $[0, \infty[$), and putting $\alpha = \inf \operatorname{supp}(\kappa)$ we get that

$$\alpha = a + b \quad \text{with } a \in \operatorname{supp}(\kappa) \text{ and } b \in \operatorname{supp}(\sigma_\varepsilon),$$

which is impossible since $a \geqq \alpha$, $b \geqq \varepsilon > 0$, and it follows that $\kappa * \sigma_\varepsilon \neq \kappa$. The measures $\kappa * \sigma_\varepsilon$ and κ are both zero on $]-\infty, 0[$, so (12) shows that $\kappa * \sigma_\varepsilon$ and κ are equal on $\mathbb{R} \setminus [-\varepsilon, \varepsilon]$.

The condition (iii) of Definition 17.1 is satisfied because

$$\operatorname{supp}(\sigma_\varepsilon^n * \kappa) \subseteq [n\varepsilon, \infty[\quad \text{for } n \in \mathbb{N}.$$

In order to prove (iv) of 17.1 we first prove that if σ is a positive measure with compact support contained in $[0, \infty[$ such that

$$\kappa * \sigma \leqq \kappa,$$

then $\sigma([0, \infty[) \leqq 1$.

Let η be the positive measure supported by $[0, \infty[$ such that $\kappa = \kappa * \sigma + \eta$. The Laplace transform of κ is well-defined because κ is shift-bounded. In fact, for $x > 0$ we have

$$\int_0^\infty e^{-xs} d\kappa(s) \leqq \sum_{n=0}^\infty e^{-xn} \kappa([n, n+1]) < \infty,$$

because $\sup_{n\in\mathbb{N}} \kappa([n, n+1]) < \infty$.

It follows that

$$\mathscr{L}\kappa(x) = \mathscr{L}\kappa(x)\mathscr{L}\sigma(x) + \mathscr{L}\eta(x) \quad \text{for } x>0,$$

and hence

$$\mathscr{L}\sigma(x) \leqq 1 \quad \text{for } x>0,$$

which by 9.4 (2) shows that $\sigma([0, \infty[) \leqq 1$.

By (12) we now find that $\sigma_\varepsilon(K) \leqq 1$ for every compact subset K of $[0, \infty[$, hence $\sigma_\varepsilon([0, \infty[) \leqq 1$. This shows that κ is a perfect kernel and by Theorem 17.8 it follows that κ is potential kernel for a transient convolution semigroup $(\mu_t)_{t>0}$ on $\mathbb{R}$. Clearly $\operatorname{supp}(\mu_t) \subseteq \operatorname{supp}(\kappa)$, so the semigroup $(\mu_t)_{t>0}$ is supported by $[0, \infty[$. □

17.12. Remark. Denoting by a_ε, for $\varepsilon \in]0, 1[$, the positive number such that the measure

$$\eta_\varepsilon = a_\varepsilon \kappa * (\varepsilon_0 - \sigma_\varepsilon)$$

has total mass 1, we get for the Laplace transforms

$$\frac{\mathscr{L}\eta_\varepsilon(x)}{\mathscr{L}\kappa(x)} = a_\varepsilon(1 - \mathscr{L}\sigma_\varepsilon(x)) \quad \text{for } x>0,$$

from which it easily follows that $f = \dfrac{1}{\mathscr{L}\kappa}$ is a Bernstein function. This shows by Proposition 14.1 that κ is potential kernel for the transient convolution semigroup on $\mathbb{R}$ supported by $[0, \infty[$ with corresponding Bernstein function f. It is therefore not necessary to use Theorem 17.8 in the proof of Theorem 17.11.

Let $(\sigma_V)_{V \in \check{\mathscr{V}}}$ be a fundamental family associated with the perfect kernel κ and let $(\mu_t)_{t>0}$ be the uniquely determined transient convolution semigroup on G with potential kernel κ.

17.13. Proposition. *A positive measure ξ on G is excessive (resp. invariant) with respect to $(\mu_t)_{t>0}$ if and only if ξ is σ_V-superharmonic (resp. σ_V-harmonic) for all $V \in \check{\mathscr{V}}$, i.e. if and only if $\xi \in D^+(\sigma_V)$ and $\sigma_V * \xi \leqq \xi$ (resp. $\sigma_V * \xi = \xi$) for all $V \in \check{\mathscr{V}}$.*

Proof. The proof will be accomplished in the following four steps:

1°. If ξ is σ_V-superharmonic for all $V \in \check{\mathscr{V}}$ then ξ is excessive.

Suppose that ξ is σ_V-superharmonic for all $V \in \check{\mathscr{V}}$, and let $t>0$ be fixed. Using the notations of 17.6 we find for all $n \geqq 0$ and $V \in \check{\mathscr{V}}$ that

$$(t a_V \sigma_V)^n * \xi \leqq (t a_V)^n \xi,$$

and therefore

$$e^{-t a_V} \sum_{n=0}^{N} \frac{1}{n!} (t a_V \sigma_V)^n * \xi \leqq e^{-t a_V} \left(\sum_{n=0}^{N} \frac{1}{n!} (t a_V)^n \right) \xi \leqq \xi$$

for all $N \in \mathbb{N}$ and $V \in \check{V}$. Letting N tend to infinity we get by (11) and Lemma 15.3 that $\xi \in D^+(\mu_t^V)$ and $\mu_t^V * \xi \leqq \xi$ for $V \in \check{V}$.

By Proposition 17.9 we have in particular that

$$\lim_{\check{V}} \mu_t^V = \mu_t \text{ vaguely,}$$

and by Lemma 15.1 we therefore get that $\xi \in D^+(\mu_t)$ and $\mu_t * \xi \leqq \xi$, which shows that ξ is excessive.

2°. If ξ is excessive then ξ is σ_V-superharmonic for all $V \in \check{V}$.

Let $V \in \check{V}$ be fixed. By Proposition 16.29 and Proposition 15.16 it suffices to prove that every κ-potential $\kappa * \sigma$, with $\sigma \in D^+(\kappa)$, is σ_V-superharmonic, and this follows from the inequality

$$\sigma_V * \kappa \leqq \kappa$$

by convolution with σ.

3°. If ξ is σ_V-harmonic for $V \in \check{V}$ then ξ is invariant.

Suppose that ξ is σ_V-harmonic for all $V \in \check{V}$. In particular ξ is σ_V-superharmonic and hence excessive by 1°. By the Riesz decomposition theorem (16.7) ξ can be written

$$\xi = \kappa * \sigma + \eta,$$

where $\sigma \in D^+(\kappa)$ and η is invariant. For $V \in \check{V}$ we find, using that (by 2°) η is σ_V-superharmonic

$$\xi = \sigma_V * \xi = \sigma_V * (\kappa * \sigma) + \sigma_V * \eta \leqq \kappa * \sigma + \eta = \xi,$$

and therefore

$$\sigma_V * (\kappa * \sigma) = \kappa * \sigma.$$

Using that $\eta_V = a_V(\kappa - \sigma_V * \kappa)$, cf. 17.6, this gives that

$$\eta_V * \sigma = 0,$$

and it follows that $\sigma = 0$ (because $\eta_V \neq 0$), and that $\xi = \eta$ is invariant.

4°. If ξ is invariant then ξ is σ_V-harmonic for all $V \in \check{V}$.

Suppose that ξ is invariant and let $V \in \check{V}$ be fixed. In particular ξ is excessive and therefore σ_V-superharmonic by 2°.

It follows by (4) that the elementary kernel determined by σ_V exists and is equal to $a_V \kappa_V$, and by Proposition 15.18 we can therefore write

$$\xi = a_V \kappa_V * \sigma + \eta,$$

where $\sigma \in D^+(\kappa_V)$ and η is σ_V-harmonic.

The measure η is equal to the vague limit of the decreasing sequence $(\sigma_V^n * \xi)_{n \in \mathbb{N}}$ which consists of invariant measures, and it follows by Exercise 15.4 that η is invariant.

Using (5) we get

$$\xi * \eta_V = a_V \kappa * \sigma + \eta * \eta_V,$$

and by convolution with μ_t for a $t>0$ we obtain

$$\mu_t * (\xi * \eta_V) = \xi * \eta_V = a_V \mu_t * (\kappa * \sigma) + \mu_t * (\eta * \eta_V)$$
$$\leqq a_V \kappa * \sigma + \eta * \eta_V = \xi * \eta_V,$$

and it follows that

$$a_V \mu_t * (\kappa * \sigma) = a_V (\kappa * \sigma),$$

or

$$(\kappa - \mu_t * \kappa) * \sigma = 0.$$

Since $\kappa - \mu_t * \kappa$ is a non-zero positive measure, it follows that $\sigma = 0$, hence that $\xi = \eta$ is σ_V-harmonic. □

17.14. Remark. Let κ be a perfect kernel and let $(\sigma_V)_{V \in \mathring{V}}$ and $(\sigma'_W)_{W \in \mathring{W}}$ be fundamental families associated with κ. It follows by Proposition 17.13 that a positive measure ξ on G is σ_V-superharmonic (resp. σ_V-harmonic) for all $V \in \mathring{V}$ if and only if ξ is σ'_W-superharmonic (resp. σ'_W-harmonic) for all $W \in \mathring{W}$. This is useful in the study of concrete examples.

17.15. Example. Let κ be the elementary kernel on G determined by a positive measure μ on G such that $\mu(G) \leqq 1$, i.e.

$$\kappa = \sum_{n=0}^{\infty} \mu^n.$$

As remarked in 17.3, κ is a perfect kernel and the constant net $(\mu)_{K \in \mathring{\mathfrak{K}}(0)}$ is a fundamental family associated with κ. The transient convolution semigroup $(\mu_t)_{t>0}$ on G, for which κ is potential kernel, is the convolution semigroup determined by μ (cf. 8.10 and 13.6) i.e.

$$\mu_t = e^{-t} \exp(t\mu) \quad \text{for } t>0.$$

A positive measure ξ on G is excessive (resp. invariant) with respect to $(\mu_t)_{t>0}$ if and only if ξ is μ-superharmonic (resp. μ-harmonic).

17.16. Example. Let $G = \mathbb{R}^n$, $n \geqq 3$, and let κ_n be the Newtonian kernel, i.e.

$$\kappa_n = k_n N_n(x)\, dx,$$

cf. 13.29. The Newtonian kernel is perfect since it is the potential kernel of the Brownian semigroup.

Let V_r for $r>0$ be the compact ball

$$V_r = \{x \in \mathbb{R}^n \mid \|x\| \leqq r\},$$

and let σ_r denote the normalized surface measure for the boundary of V_r. The following formula is well-known (cf. e.g. Brelot [1])

$$\sigma_r * N_n(x) = \begin{cases} N_n(x) & \text{for } \|x\| > r, \\ r^{2-n} & \text{for } \|x\| \leqq r, \end{cases}$$

and this shows that σ_r is a κ_n-balayaged measure of ε_0 on the open set $\complement V_r$, and the net $(\sigma_r)_{r>0}$ is consequently a fundamental family associated with κ_n.

17.17. Exercise. Let $h_n\colon \mathbb{R}^n \to [0, \infty[$, $n \geqq 3$, be defined as

$$h_n(x) = \begin{cases} 1 & \text{for } \|x\| \leqq 1, \\ \|x\|^{2-n} & \text{for } \|x\| > 1, \end{cases}$$

and let κ_n be the Newtonian kernel. Then a positive measure σ on $\mathbb{R}^n$ belongs to $D^+(\kappa_n)$ if and only if

$$\int h_n(x)\, d\sigma(x) < \infty.$$

17.18. Exercise. Let $G = \mathbb{R}$. The potential kernel for the translation semigroup $(\varepsilon_t)_{t>0}$, cf. 13.9,

$$\kappa = 1_{]0,\infty[}(x)\, dx,$$

is a perfect kernel. The net $(\varepsilon_r)_{r>0}$ is a fundamental family associated with κ, where $r>0$ corresponds to the compact neighbourhood $[-r, r]$ of 0.

A positive measure σ on $\mathbb{R}$ belongs to $D^+(\kappa)$ if and only if

$$\sigma(]-\infty, x[) < \infty \quad \text{for all } x \in \mathbb{R}.$$

17.19. Exercise. For every $a>0$ the measure

$$\kappa = e^{-a|x|}\, dx$$

on $\mathbb{R}$ is a perfect kernel. Defining for $r>0$ the measure

$$\sigma_r = \frac{1}{2\cosh(ar)}(\varepsilon_r + \varepsilon_{-r}),$$

the net $(\sigma_r)_{r>0}$ is a fundamental family associated with κ, where $r>0$ corresponds to the compact neighbourhood $[-r, r]$ of 0. The measure $\frac{1}{2a}\kappa$ is equal to the resolvent measure ρ_{a^2} for the Brownian semigroup $(\mu_t)_{t>0}$ in one dimension, i.e. $\frac{1}{2a}\kappa$ is the potential kernel for the convolution semigroup

$$(e^{-a^2 t}\mu_t)_{t>0}.$$

Analogous results are valid in higher dimensions. The general formulas involve Bessel functions.

17.20. Example. Let $G=\mathbb{R}^n$ and let $\kappa_{n,\alpha}$ be the Riesz kernel of order α, cf. 14.30, which is a perfect kernel.

For $r>0$ (corresponding to the compact ball $V_r=\{x\in\mathbb{R}^n \mid \|x\|\leq r\}$) we consider the following function on $\mathbb{R}^n$

$$x\mapsto\begin{cases}\Gamma\left(\frac{n}{2}\right)\pi^{-\frac{n}{2}-1}\sin\left(\frac{\pi\alpha}{2}\right)r^{\alpha}(\|x\|^2-r^2)^{-\frac{\alpha}{2}}\|x\|^{-n} & \text{for } \|x\|>r,\\ \infty & \text{for } \|x\|=r,\\ 0 & \text{for } \|x\|<r,\end{cases}$$

which is density for a positive measure $\sigma_{r,\alpha}$ on $\mathbb{R}^n$ with total mass 1 and such that $\operatorname{supp}(\sigma_{r,\alpha})\subseteq \complement V_r$. The net $(\sigma_{r,\alpha})_{r>0}$ is a fundamental family associated with $\kappa_{n,\alpha}$, cf. Landkoff [1] p. 112.

17.21. The notion of a fundamental family associated with a convolution kernel is due to Deny [1]. Choquet and Deny [2] announced the Theorems 17.4 and 17.8 and Deny [5] proved a more general result. The perfect kernels as defined here form, on account of the condition (iv), a more restrictive class than the perfect kernels defined by Choquet and Deny [3]. The perfect kernels considered here are exactly the perfect kernels in the sense of Choquet and Deny which are shift-bounded, cf. Lemma 17.7 and Choquet and Deny [3], Théorème 27.

§ 18. The Lévy Measure for a Convolution Semigroup

The purpose of this paragraph is to describe the infinitesimal generator A for the contraction semigroups induced by a convolution semigroup $(\mu_t)_{t>0}$ on the LCA-group G. One of the main tools in this description is a certain measure on $G\setminus\{0\}$, the Lévy measure for $(\mu_t)_{t>0}$, and we shall start with a construction of this measure. The Lévy measure can be characterized in several ways, using the fact that the domain of A is a "rich" set of functions.

Then we will discuss the role of the Lévy measure in the Lévy-Khinchin representation of the continuous negative definite function on the dual group Γ associated with $(\mu_t)_{t>0}$.

A convolution semigroup $(\mu_t)_{t>0}$ on G is said to be of local type if its infinitesimal generator is a local operator, and this is shown to be the case if and only if the Lévy measure for $(\mu_t)_{t>0}$ vanishes. As a simple consequence, the general form of a continuous negative definite function associated with a convolution semigroup of local type can be given.

Finally, for a transient convolution semigroup with potential kernel κ, the Lévy measure and the notion of locality will be expressed in terms of the fundamental families associated with κ.

Let S denote the set of symmetric probability measures on Γ with compact support, i.e.

$$S=\{\sigma\in M_1^+(\Gamma)\cap M_c(\Gamma)\,|\,\sigma=\check{\sigma}\}.$$

The restriction of a measure μ on G to a Borel subset $B\subseteq G$ will be denoted $\mu|B$.

18.1. Lemma. *For every compact neighbourhood V of 0 in G there exists a measure $\sigma\in S$ such that*

$$\mathscr{F}_\Gamma\,\sigma(x)\leqq\tfrac{1}{2}\quad \textit{for } x\in G\setminus V.$$

Proof. There exists a function $\varphi\in CP(G)$ such that

$$0\leqq\varphi\leqq 1,\quad \varphi(0)=1 \quad\text{and}\quad \operatorname{supp}(\varphi)\subseteq V.$$

The associated measure ν on Γ (cf. 3.12) is a symmetric probability measure and by choosing a compact symmetric set $K\subseteq\Gamma$ such that $\nu(K)\geqq\frac{3}{4}$ and putting $\sigma=\nu(K)^{-1}(\nu|K)$ we find

$$\|\varphi-\mathscr{F}_\Gamma\sigma\|_\infty\leqq\|\nu-\sigma\|\leqq\tfrac{1}{2}$$

and hence

$$\mathscr{F}_\Gamma\,\sigma(x)\leqq\tfrac{1}{2}\quad\text{for } x\in G\setminus V.\quad\square$$

18.2. Proposition. *Let $(\mu_t)_{t>0}$ be a convolution semigroup on G with associated continuous negative definite function ψ on Γ. The net $\left(\frac{1}{t}\mu_t|G\setminus\{0\}\right)_{t>0}$ of positive measures on $G\setminus\{0\}$ converges vaguely as $t\to 0$ to a measure μ on $G\setminus\{0\}$. For every $\sigma\in S$ the function $\psi*\sigma-\psi$ is continuous and positive definite on Γ and the positive bounded measure μ_σ on G whose Fourier transform is $\psi*\sigma-\psi$, satisfies*

$$(1-\mathscr{F}_\Gamma\sigma)\mu=\mu_\sigma|G\setminus\{0\}. \tag{1}$$

Proof. Let $\sigma\in S$. The measure $(1-\mathscr{F}_\Gamma\sigma)\frac{1}{t}\mu_t$ is for $t>0$ a positive bounded measure on G whose Fourier transform is given by

$$\begin{aligned}
\mathscr{F}_G\left[(1-\mathscr{F}_\Gamma\sigma)\frac{1}{t}\mu_t\right](\gamma)&=\frac{1}{t}\int\overline{(x,\gamma)}(1-\mathscr{F}_\Gamma\sigma(x))\,d\mu_t(x)\\
&=\frac{1}{t}\int\overline{(x,\gamma)}\left[1-\int(x,\delta)\,d\sigma(\delta)\right]d\mu_t(x)\\
&=\frac{1}{t}\left[\hat{\mu}_t(\gamma)-\iint\overline{(x,\gamma-\delta)}\,d\sigma(\delta)\,d\mu_t(x)\right]\\
&=\frac{1}{t}\left[\hat{\mu}_t(\gamma)-\hat{\mu}_t*\sigma(\gamma)\right]\\
&=\frac{1}{t}\left[1-\exp(-t\psi)\right]*(\sigma-\varepsilon_0)(\gamma)
\end{aligned}$$

for $\gamma \in \Gamma$. Since $\lim_{t \to 0} \frac{1}{t}(1 - \exp(-t\psi)) = \psi$, uniformly on compact subsets of Γ, we find that

$$\lim_{t \to 0} \mathscr{F}_G \left[(1 - \mathscr{F}_\Gamma \sigma) \frac{1}{t} \mu_t \right] (\gamma) = \psi * \sigma(\gamma) - \psi(\gamma),$$

pointwise (or uniformly over compact sets) on Γ. This shows that the function $\gamma \mapsto \psi * \sigma(\gamma) - \psi(\gamma)$ is (continuous and) positive definite, and furthermore (by 3.13) that

$$\lim_{t \to 0} (1 - \mathscr{F}_\Gamma \sigma) \frac{1}{t} \mu_t = \mu_\sigma$$

in the Bernoulli topology on G, where μ_σ is the positive bounded measure on G such that

$$\hat{\mu}_\sigma = \psi * \sigma - \psi .$$

For $\varphi \in C_c^+(G)$ satisfying $\operatorname{supp}(\varphi) \subseteq G \setminus \{0\}$ we may choose, by Lemma 18.1, $\sigma \in S$ such that $\mathscr{F}_\Gamma \sigma \leqq \frac{1}{2}$ in a neighbourhood of $\operatorname{supp}(\varphi)$, and the function

$$x \overset{\varphi'}{\longmapsto} \begin{cases} \dfrac{\varphi(x)}{1 - \mathscr{F}_\Gamma \sigma(x)} & \text{for } x \in \operatorname{supp}(\varphi), \\ 0 & \text{for } x \notin \operatorname{supp}(\varphi), \end{cases}$$

then belongs to $C_c^+(G)$, and it follows that

$$\lim_{t \to 0} \langle \frac{1}{t} \mu_t, \varphi \rangle = \lim_{t \to 0} \langle (1 - \mathscr{F}_\Gamma \sigma) \frac{1}{t} \mu_t, \varphi' \rangle = \langle \mu_\sigma, \varphi' \rangle .$$

This shows, that there exists a positive measure μ on $G \setminus \{0\}$ such that

$$\mu = \lim_{t \to 0} \frac{1}{t} \mu_t | G \setminus \{0\} \qquad \text{vaguely on } G \setminus \{0\},$$

and furthermore that the measure μ satisfies

$$(1 - \mathscr{F}_\Gamma \sigma) \mu = \mu_\sigma | G \setminus \{0\} \qquad \text{for } \sigma \in S . \quad \square$$

18.3. Definition. The positive measure μ on $G \setminus \{0\}$ from Proposition 18.2 is called the *Lévy measure* for the convolution semigroup $(\mu_t)_{t>0}$ on G (and also the Lévy measure for the continuous negative definite function ψ on Γ).

18.4. Proposition. *Let μ be the Lévy measure for the convolution semigroup $(\mu_t)_{t>0}$ on G. Then*

$$\int_{G \setminus \{0\}} [1 - \operatorname{Re}(x, \gamma)] \, d\mu(x) < \infty \qquad \textit{for all } \gamma \in \Gamma,$$

and the restriction of μ to the complement of a compact neighbourhood V of 0 is a bounded measure on $G \setminus V$ (and on G).

Proof. The first assertion follows from the characterization (1) of the Lévy measure by taking $\sigma=\frac{1}{2}(\varepsilon_\gamma+\varepsilon_{-\gamma})$ for $\gamma\in\Gamma$. Let V be a compact neighbourhood of 0. There exists a measure $\sigma\in S$ such that

$$1-\mathscr{F}_\Gamma\sigma(x)\geqq\frac{1}{2}\quad\text{for } x\in G\setminus V,$$

(cf. Lemma 18.1), and it follows by (1) that

$$\langle\mu,\varphi\rangle\leqq 2\langle(1-\mathscr{F}_\Gamma\sigma)\mu,\varphi\rangle=2\langle\mu_\sigma,\varphi\rangle$$

for all $\varphi\in C_c^+(G)$ with $\operatorname{supp}(\varphi)\subseteq G\setminus V$, and the result follows since μ_σ is a bounded measure on G. □

18.5. Exercise. The Lévy measure for the convolution semigroup $(e^{-t}\exp(t\mu))_{t>0}$ on G, determined by the positive bounded measure μ on G with total mass $\mu(G)\leqq 1$, is $\mu\,|\,G\setminus\{0\}$.

18.6. Exercise. Let $(\mu_t')_{t>0}$ and $(\mu_t'')_{t>0}$ be convolution semigroups on G with Lévy measures μ' and μ''. The Lévy measure for the "mixed" convolution semigroup $(\mu_t'*\mu_t'')_{t>0}$ is $\mu'+\mu''$.

18.7. Proposition. *Let $(\eta_t)_{t>0}$ be a convolution semigroup on $\mathbb{R}$ supported by $[0,\infty[$, and suppose that the corresponding Bernstein function f has the representation (cf. Theorem 9.8)*

$$f(x)=a+bx+\int_0^\infty(1-e^{-xs})\,d\nu(s)\quad\textit{for } x>0.$$

The Lévy measure for $(\eta_t)_{t>0}$ is ν.

Proof. The associated continuous, negative definite function on $\hat{\mathbb{R}}\approx\mathbb{R}$ is given by

$$\psi(y)=a+iby+\int_0^\infty(1-e^{-iys})\,d\nu(s)\quad\text{for } y\in\mathbb{R},$$

(cf. 9.19) and it follows that for every symmetric probability measure σ on $\hat{\mathbb{R}}\approx\mathbb{R}$ with compact support we have

$$\begin{aligned}\psi*\sigma(y)-\psi(y)&=\int_0^\infty(\varepsilon_0-\sigma)*[x\mapsto e^{-ixs}](y)\,d\nu(s)\\&=\int_0^\infty e^{-iys}(1-\mathscr{F}_{\mathbb{R}}\sigma(s))\,d\nu(s)\\&=\mathscr{F}_{\mathbb{R}}[(1-\mathscr{F}_{\mathbb{R}}\sigma)\nu](y),\end{aligned}$$

and the assertion now follows from the characterization (1) of the Lévy measure. □

18.8. Proposition. *Let $(\mu_t)_{t>0}$ be a convolution semigroup on G with Lévy measure μ and let $(\eta_t)_{t>0}$ be a convolution semigroup on $\mathbb{R}$ supported by $[0,\infty[$ with corresponding Bernstein function f given by*

$$f(x)=a+bx+\int_0^\infty (1-e^{-xs})\,d\nu(s) \quad \textit{for } x>0.$$

$(\mu_t)_{t>0}$ by means of $(\eta_t)_{t>0}$, cf. 9.20, is given by

$$\mu^f=b\mu+\int_0^\infty [\mu_s \mid G\setminus\{0\}]\,d\nu(s) \quad \textit{vaguely.}$$

Proof. The continuous, negative definite function on Γ associated with $(\mu_t^f)_{t>0}$ is the function $f\circ\psi$ (cf. 9.20), i.e.

$$f\circ\psi(\gamma)=a+b\psi(\gamma)+\int_0^\infty (1-e^{-\psi(\gamma)s})\,d\nu(s) \qquad \text{for } \gamma\in\Gamma,$$

where ψ is the continuous negative definite function associated with $(\mu_t)_{t>0}$. Let $\sigma\in S$. By Proposition 18.2 there exist positive bounded measures μ_σ and μ_σ^f on G such that

$$\psi*\sigma-\psi=\mathscr{F}_G\mu_\sigma \quad \text{and} \quad (f\circ\psi)*\sigma-f\circ\psi=\mathscr{F}_G\mu_\sigma^f,$$

and we find for $\gamma\in\Gamma$ that

$$\begin{aligned}\mathscr{F}_G\mu_\sigma^f(\gamma)&=(f\circ\psi)*\sigma(\gamma)-f\circ\psi(\gamma)\\ &=b(\psi*\sigma(\gamma)-\psi(\gamma))+\int_0^\infty e^{-s\psi}*(\varepsilon_0-\sigma)(\gamma)\,d\nu(s)\\ &=b\mathscr{F}_G\mu_\sigma(\gamma)+\int_0^\infty \mathscr{F}_G((1-\mathscr{F}_\Gamma\sigma)\mu_s)(\gamma)\,d\nu(s).\end{aligned}$$

From the proof of Proposition 18.2 we know that

$$\lim_{s\to 0}(1-\mathscr{F}_\Gamma\sigma)\frac{1}{s}\mu_s=\mu_\sigma$$

in the Bernoulli topology, so for every function $g\in C_b(G)$ there exists a constant $c>0$ such that

$$|\langle\mu_s,(1-\mathscr{F}_\Gamma\sigma)g\rangle|\leqq cs \qquad \text{for } s\in\,]0,1[,$$

and it follows from Theorem 9.8 that the integral

$$\int_0^\infty \langle\mu_s,(1-\mathscr{F}_\Gamma\sigma)g\rangle\,d\nu(s)$$

exists for all $g \in C_b(G)$. The vague integral

$$\tau_\sigma = \int_0^\infty (1 - \mathscr{F}_\Gamma \sigma)\mu_s \, dv(s)$$

defines consequently a positive bounded measure on G with Fourier transform

$$\mathscr{F}_G \tau_\sigma(\gamma) = \int_0^\infty \langle \mu_s, (1 - \mathscr{F}_\Gamma \sigma)\bar{\gamma}\rangle \, dv(s) = \int_0^\infty \mathscr{F}_G((1 - \mathscr{F}_\Gamma \sigma)\mu_s)(\gamma) \, dv(s),$$

and we therefore have

$$\mu_\sigma^f = b\mu_\sigma + \tau_\sigma.$$

Let $\varphi \in C_c(G \setminus \{0\})$. There exists a measure $\sigma_0 \in S$ such that $\mathscr{F}_\Gamma \sigma_0(x) \leqq \frac{1}{2}$ for $x \in \operatorname{supp}(\varphi)$ (cf. 18.1), and applying the formula $\mu_{\sigma_0}^f = b\mu_{\sigma_0} + \tau_{\sigma_0}$ to the function

$$x \mapsto \begin{cases} \dfrac{\varphi(x)}{1 - \mathscr{F}_\Gamma \sigma_0(x)} & \text{for } x \in \operatorname{supp}(\varphi), \\ 0 & \text{for } x \notin \operatorname{supp}(\varphi), \end{cases}$$

we find by (1)

$$\langle \mu^f, \varphi \rangle = b\langle \mu, \varphi \rangle + \int_0^\infty \langle \mu_s, \varphi \rangle \, dv(s),$$

which proves the asserted formula for the Lévy measure μ^f for the convolution semigroup $(\mu_t^f)_{t>0}$. □

Let $(\mu_t)_{t>0}$ be a convolution semigroup on G with resolvent $(\rho_\lambda)_{\lambda>0}$ and with associated continuous negative definite function ψ on Γ. The strongly continuous contraction semigroup on $C_0(G)$ or $L^p(G)$ for $p \in [1, \infty[$, induced by $(\mu_t)_{t>0}$ (cf. 12.7 and 12.9), is denoted $(P_t)_{t>0}$. The infinitesimal generator for $(P_t)_{t>0}$ on $C_0(G)$ (resp. on $L^p(G)$ for $p \in [1, \infty[$) is denoted $(A_0, D(A_0))$ (resp. $(A_p, D(A_p))$).

18.9. Proposition. *The set $C_c(G) \cap D(A_0) \bigcap_{p \geqq 1} D(A_p)$ is dense in $C_c(G)$ with the inductive limit topology and dense in $C_0(G)$ and $L^p(G)$ for $p \in [1, \infty[$.*

Proof. Let $f \in C_c^+(G)$ and $\varepsilon > 0$ be given. We shall see that there exists for every neighbourhood U of 0 a function $g \in C_c^+(G) \cap D(A_0) \bigcap_{p \geqq 1} D(A_p)$ such that

$$\|f - g\|_\infty < \varepsilon \quad \text{and} \quad \operatorname{supp}(g) \subseteq \operatorname{supp}(f) + U. \tag{2}$$

Since f is uniformly continuous, there exists a compact neighbourhood V of 0 such that $V \subseteq U$ and

$$|f(x-y) - f(x)| \leqq \varepsilon \quad \text{for } x \in G \text{ and } y \in V.$$

The measure ρ_1 is potential kernel for the transient convolution semigroup $(e^{-t}\mu_t)_{t>0}$, and it follows that ρ_1 is a perfect kernel (cf. Theorem 17.4). In particular there exists a positive bounded measure σ_V on G with total mass $\sigma_V(G) \leqq 1$ and with the properties

$$\rho_1 * \sigma_V \leqq \rho_1, \quad \rho_1 * \sigma_V = \rho_1 \text{ in } \complement V \quad \text{and} \quad \rho_1 * \sigma_V \neq \rho_1.$$

For a suitable constant $a_V > 0$ the measure

$$\alpha_V = a_V(\rho_1 - \rho_1 * \sigma_V)$$

is thus a probability measure with $\operatorname{supp}(\alpha_V) \subseteq V$, and it is clear that the function

$$g = f * \alpha_V = \rho_1 * (a_V f) - \rho_1 * \sigma_V * (a_V f)$$

satisfies (2). The function $\sigma_V * (a_V f)$ belongs to $C_0(G)$ and $\bigcap_{p \geqq 1} L^p(G)$, since σ_V is bounded, and it follows (cf. Proposition 11.10) that

$$g \in \rho_1 * [C_0(G) \bigcap_{p \geqq 1} L^p(G)] \subseteq D(A_0) \bigcap_{p \geqq 1} D(A_p).$$

The rest now follows immediately. □

18.10. Corollary. *Let U and V be open, relatively compact subsets of G such that $\bar{U} \subseteq V$. There exists a function $f \in D(A_0) \bigcap_{p \geqq 1} D(A_p)$ satisfying*

$$0 \leqq f \leqq 1, \quad f = 1 \text{ in } U \quad \text{and} \quad f = 0 \text{ in } \complement V. \tag{3}$$

Proof. Let W be an open, relatively compact neighbourhood of 0 such that

$$(\bar{W} + \bar{U}) \cap (\bar{W} + \complement V) = \emptyset.$$

There exists a function $g \in C_c^+(G)$ with the properties

$$0 \leqq g \leqq 1, \quad g = 1 \text{ in } W + U \quad \text{and} \quad g = 0 \text{ in } W + \complement V.$$

For $h \in C_c^+(G) \cap D(A_0) \bigcap_{p \geqq 1} D(A_p)$ such that

$$\operatorname{supp}(h) \subseteq W \quad \text{and} \quad \int h(x)\,dx = 1,$$

the function $f = g * h$ belongs to $D(A_0) \bigcap_{p \geqq 1} D(A_p)$ (cf. 12.11) and satisfies (3). □

18.11. Remark. The Lévy measure μ for the convolution semigroup $(\mu_t)_{t>0}$ on G can be expressed in terms of the infinitesimal generators A_0 and A_p, $p \geqq 1$.

For $f \in C_c(G)$ such that $\check{f} \in D(A_0)$ and $\operatorname{supp}(f) \subseteq G \setminus \{0\}$ we have

$$\begin{aligned}\langle \mu, f \rangle &= \lim_{t \to 0} \langle \frac{1}{t} \mu_t, f \rangle = \lim_{t \to 0} \langle \frac{1}{t} (\mu_t - \varepsilon_0), f \rangle \\ &= \lim_{t \to 0} \frac{1}{t} (\mu_t - \varepsilon_0) * \check{f}(0) = A_0 \check{f}(0).\end{aligned}$$

For $f, g \in C_c(G)$ such that $f \in D(A_p)$ and $\operatorname{supp}(\check{f} * \bar{g}) \subseteq G \setminus \{0\}$ we find

$$\langle \mu, \check{f} * \bar{g} \rangle = \lim_{t \to 0} \langle \frac{1}{t}(\mu_t - \varepsilon_0) * f, \bar{g} \rangle = \langle A_p f, \bar{g} \rangle,$$

and for $p=2$ we have furthermore

$$\langle \mu, \check{f} * \bar{g} \rangle = -(\psi \hat{f}, \hat{g}),$$

where $(\,,)$ denotes the inner product in $L^2(\Gamma)$.

It is easy to see that each of the formulas

$$\langle \mu, f \rangle = A_0 \check{f}(0), \tag{4}$$

and

$$\langle \mu, \check{f} * \bar{g} \rangle = \langle A_p f, \bar{g} \rangle, \tag{5}$$

with f and g as specified above, determines the measure μ on $G \setminus \{0\}$, and (4) or (5) can even be used as starting point in the construction of μ.

18.12. Exercise. The Lévy measure μ for $(\mu_t)_{t>0}$ is vague limit for the net $(\lambda^2 \rho_\lambda \mid G \setminus \{0\})_{\lambda > 0}$ of measures on $G \setminus \{0\}$ as λ tends to ∞, where $(\rho_\lambda)_{\lambda>0}$ is the resolvent for $(\mu_t)_{t>0}$.

The Lévy-Khinchin representation of a continuous negative definite function ψ on Γ gives a decomposition of ψ as a sum of "simple" negative definite functions on Γ, and we shall start by studying these parts in terms of their Lévy measures and of the Lévy measure μ for ψ.

From the characterization (1) of the Lévy measure, it is clear that the Lévy measure for a constant negative definite function is zero.

18.13. Lemma. *A continuous function $l\colon \Gamma \to \mathbb{R}$ such that $l(0)=0$ is a homomorphism if and only if $l * \sigma - l = 0$ for all $\sigma \in S$.*

Proof. If l is a homomorphism then we have for $\sigma \in S$

$$\int l(\eta)\, d\sigma(\eta) = \int l(-\eta)\, d\check{\sigma}(\eta) = -\int l(\eta)\, d\sigma(\eta)$$

so that $\langle \sigma, l \rangle = 0$, and hence for $\gamma \in \Gamma$

$$l * \sigma(\gamma) = \int l(\gamma - \eta)\, d\sigma(\eta) = l(\gamma) - \int l(\eta)\, d\sigma(\eta) = l(\gamma).$$

If conversely $l * \sigma - l = 0$ for $\sigma \in S$, then we find for $\gamma, \eta \in \Gamma$ and $\sigma = \frac{1}{2}(\varepsilon_\gamma + \varepsilon_{-\gamma})$ that

$$0 = l * \sigma(\eta) - l(\eta) = \frac{1}{2}[l(\eta - \gamma) + l(\eta + \gamma)] - l(\eta),$$

and in particular for $\eta=0$

$$\frac{1}{2}[l(-\gamma)+l(\gamma)]=0.$$

Analogously we get

$$0=\frac{1}{2}[l(\gamma-\eta)+l(\gamma+\eta)]-l(\gamma)$$

and hence by addition that

$$l(\gamma+\eta)=l(\gamma)+l(\eta). \quad \square$$

18.14. Corollary. *The Lévy measure for a continuous, negative definite function ψ of the form $\psi=il$, where $l\colon \Gamma\to\mathbb{R}$ is a continuous homomorphism, is zero.*

Proof. Such a function is negative definite by Proposition 7.20 and the assertion follows immediately from Lemma 18.13 and the characterization (1) of the Lévy measure. $\square$

18.15. Corollary. *Let ψ be a continuous, negative definite function on Γ with Lévy measure μ. The function $i\operatorname{Im}\psi$ is negative definite if and only if μ is symmetric.*

Proof. Putting $l=\operatorname{Im}\psi$ we have that $l(0)=0$. By Proposition 7.20 the function $i\operatorname{Im}\psi$ is negative definite if and only if l is a homomorphism, and by Lemma 18.13 this is equivalent with $\psi*\sigma-\psi$ being real for all $\sigma\in S$, which is satisfied if and only if μ_σ is symmetric for all $\sigma\in S$. It follows by (1) that μ is symmetric if and only if μ_σ is symmetric for all $\sigma\in S$. $\square$

18.16. Lemma. *A continuous symmetric function $q\colon \Gamma\to\mathbb{R}$ such that $q(0)=0$ is a quadratic form on Γ if and only if the function $q*\sigma-q$ is constant for all $\sigma\in S$. Moreover, in the affirmative case q is non-negative if and only if $q*\sigma-q\geqq 0$ for all $\sigma\in S$.*

Proof. If q is a quadratic form, then we find for $\sigma\in S$ and $\gamma\in\Gamma$,

$$\begin{aligned} q*\sigma(\gamma)&=\int q(\gamma-\eta)\,d\sigma(\eta)=\int q(\gamma+\eta)\,d\sigma(\eta)\\ &=\int \tfrac{1}{2}[q(\gamma-\eta)+q(\gamma+\eta)]\,d\sigma(\eta)=\int [q(\gamma)+q(\eta)]\,d\sigma(\eta)\\ &=q(\gamma)+\int q(\eta)\,d\sigma(\eta). \end{aligned}$$

If conversely $q*\sigma-q$ is constant for all $\sigma\in S$, then we find for $\gamma,\eta\in\Gamma$ and $\sigma=\frac{1}{2}(\varepsilon_\eta+\varepsilon_{-\eta})$ that

$$q*\sigma(\gamma)-q(\gamma)=q*\sigma(0)-q(0),$$

i.e.

$$\frac{1}{2}[q(\gamma+\eta)+q(\gamma-\eta)]-q(\gamma)=\frac{1}{2}[q(\eta)+q(-\eta)]=q(\eta). \tag{6}$$

The last assertion follows easily from (6). □

18.17. Corollary. *A real, continuous, negative definite function ψ on Γ such that $\psi(0)=0$ and with Lévy measure μ is a non-negative quadratic form if and only if $\mu=0$.*

Proof. Since ψ is a non-negative symmetric function, it follows by Lemma 18.16 that ψ is a quadratic form if and only if μ_σ is concentrated in $\{0\}$ for all $\sigma\in S$. This is however the case precisely when μ is zero. □

18.18. Lemma. *Let μ be a positive symmetric measure on $G\setminus\{0\}$ such that*

$$\int_{G\setminus\{0\}}[1-\operatorname{Re}(x,\gamma)]\,d\mu(x)<\infty \qquad \text{for } \gamma\in\Gamma. \tag{7}$$

The function $\psi_\mu\colon \Gamma\to\mathbb{R}$ defined by

$$\psi_\mu(\gamma)=\int_{G\setminus\{0\}}[1-\operatorname{Re}(x,\gamma)]\,d\mu(x) \qquad \text{for } \gamma\in\Gamma,$$

is continuous and negative definite.

The Lévy measure for ψ_μ is μ, so the measure μ is in particular uniquely determined by ψ_μ.

Proof. Let $\mathring{K}$ be the set of compact subsets $K\subseteq G$ such that $0\notin K$ and denote by μ_K for $K\in\mathring{K}$ the restriction of μ to K. The function ψ_{μ_K} is clearly continuous, and it is also negative definite, since each of the functions $\gamma\mapsto 1-\operatorname{Re}(x,\gamma)$ for $x\in G$ is negative definite (cf. 7.7).

It follows that

$$\psi_\mu=\sup_{K\in\mathring{K}}\psi_{\mu_K}$$

is lower semicontinuous and negative definite.

Moreover the function ψ_μ is locally bounded. To see this we consider for $n\in\mathbb{N}$ the set $A_n=\{\gamma\in\Gamma \mid \psi_\mu(\gamma)\leqq n\}$, which is closed. Since ψ_μ is finite everywhere we have

$$\Gamma=\bigcup_{n=1}^{\infty}A_n.$$

It follows by Baire's theorem that there exists an $n_0\in\mathbb{N}$ such that A_{n_0} has a non-empty interior. Let V be an open neighbourhood of 0 and $\gamma_1\in A_{n_0}$ be such that $\gamma_1+V\subseteq A_{n_0}$, and consider $\gamma_0\in\Gamma$. Since $\psi_\mu\in N(\Gamma)$ we find by Proposition 7.15 for $\gamma\in\gamma_0+V$ and $\gamma_2=\gamma_0-\gamma_1$ that

$$\sqrt{\psi_\mu(\gamma)}=\sqrt{\psi_\mu(\gamma-\gamma_2+\gamma_2)}\leqq\sqrt{\psi_\mu(\gamma-\gamma_2)}+\sqrt{\psi_\mu(\gamma_2)}\leqq\sqrt{\psi_\mu(\gamma_2)}+\sqrt{n_0},$$

and this shows that ψ_μ is locally bounded.

It follows that ψ_μ is locally integrable, and for $f \in C_c^+(\Gamma)$ such that $f = \check{f}$ and $\int f(\gamma)\,d\gamma = 1$ we find

$$\begin{aligned}\psi_\mu * f(\gamma) &= \int_\Gamma f(\gamma-\eta)\Big[\int_{G\setminus\{0\}} (1-\operatorname{Re}(x,\eta))\,d\mu(x)\Big]\,d\eta\\ &= \int_{G\setminus\{0\}} (1-\mathscr{F}_\Gamma f(x)\operatorname{Re}(x,\gamma))\,d\mu(x)\\ &= \psi_\mu(\gamma) + \int_{G\setminus\{0\}} \operatorname{Re}(x,\gamma)(1-\mathscr{F}_\Gamma f(x))\,d\mu(x).\end{aligned}$$

In particular we have for $\gamma = 0$

$$\int_{G\setminus\{0\}} (1-\mathscr{F}_\Gamma f(x))\,d\mu(x) = \int f(\eta)\,\psi_\mu(\eta)\,d\eta < \infty.$$

The measure $(1-\mathscr{F}_\Gamma f(x))\,d\mu(x)$, which is thus a positive bounded measure on $G\setminus\{0\}$, can be considered as a positive bounded measure τ on G, and we have that

$$\hat{\tau}(\gamma) = \operatorname{Re}\hat{\tau}(\gamma) = \int_{G\setminus\{0\}} \operatorname{Re}(x,\gamma)[1-\mathscr{F}_\Gamma f(x)]\,d\mu(x) \qquad \text{for } \gamma \in \Gamma.$$

It follows that

$$\psi_\mu = \psi_\mu * f - \hat{\tau}$$

is a continuous function on Γ.

It remains to see that μ is the Lévy measure for ψ_μ. For $\sigma \in S$ and $\gamma \in \Gamma$ we have

$$\begin{aligned}\psi_\mu * \sigma(\gamma) - \psi_\mu(\gamma) &= \int_\Gamma \Big(\int_{G\setminus\{0\}} (1-\operatorname{Re}(x,\gamma-\eta))\,d\mu(x)\Big)\,d\sigma(\eta) - \int_{G\setminus\{0\}} (1-\operatorname{Re}(x,\gamma))\,d\mu(x)\\ &= \int_{G\setminus\{0\}} \operatorname{Re}(x,\gamma)(1-\mathscr{F}_\Gamma\sigma(x))\,d\mu(x)\\ &= \int_{G\setminus\{0\}} (x,\gamma)(1-\mathscr{F}_\Gamma\sigma(x))\,d\mu(x),\end{aligned}$$

and the measure $(1-\mathscr{F}_\Gamma\sigma(x))\,d\mu(x)$ is thus (considered as a positive bounded measure on G) the measure whose Fourier transform is the function $\psi_\mu * \sigma - \psi_\mu$. The assertion now follows from the characterization (1) of the Lévy measure. □

18.19. Theorem. *A function $\psi\colon \Gamma \to \mathbb{C}$ is a continuous, negative definite function with symmetric Lévy measure if and only if it can be written*

$$\psi(\gamma) = c + il(\gamma) + q(\gamma) + \int_{G\setminus\{0\}} (1-\operatorname{Re}(x,\gamma))\,d\mu(x) \qquad \textit{for } \gamma \in \Gamma, \tag{8}$$

where c is a non-negative constant, l is a continuous homomorphism of Γ into $\mathbb{R}$, q is a non-negative, continuous quadratic form on Γ and μ is a positive, symmetric measure on $G\setminus\{0\}$ such that

$$\int_{G\setminus\{0\}} (1-\operatorname{Re}(x,\gamma))\,d\mu(x) < \infty \qquad \textit{for } \gamma \in \Gamma. \tag{9}$$

Moreover, the quadruple (c, l, q, μ) *is uniquely determined by* ψ: $c=\psi(0)$, $l=\operatorname{Im}\psi$, μ *is the Lévy measure for* ψ *and*

$$q(\gamma)=\lim_{n\to\infty}\frac{\psi(n\gamma)}{n^2} \quad \textit{for } \gamma\in\Gamma. \tag{10}$$

Proof. Suppose first that ψ is continuous and negative definite on Γ and that the Lévy measure μ for ψ is symmetric. The imaginary part $\operatorname{Im}\psi$ of ψ is by Corollary 18.15 of the form $\operatorname{Im}\psi=l$, where $l\colon\Gamma\to\mathbb{R}$ is a continuous homomorphism. Let $c=\psi(0)$. The function defined by

$$\psi'(\gamma)=\operatorname{Re}\psi(\gamma)-c \quad \text{for } \gamma\in\Gamma,$$

is continuous and negative definite (cf. Corollary 7.6), and the Lévy measure for ψ' is clearly μ.

By Proposition 18.4 the function

$$\psi_\mu(\gamma)=\int_{G\setminus\{0\}}[1-\operatorname{Re}(x,\gamma)]\,d\mu(x) \quad \text{for } \gamma\in\Gamma,$$

is finite in all points $\gamma\in\Gamma$, and by Lemma 18.18 it follows that the function

$$q=\psi'-\psi_\mu,$$

is continuous, real and symmetric and $q(0)=0$. Since $\psi'=\psi-i\operatorname{Im}\psi-c$, it is clear by Lemma 18.13 that

$$\psi'*\sigma-\psi'=\psi*\sigma-\psi \quad \text{for } \sigma\in S.$$

It follows that

$$\hat\mu_\sigma=\psi'*\sigma-\psi',$$

where μ_σ is the positive bounded measure on G such that

$$\hat\mu_\sigma=\psi*\sigma-\psi.$$

On the other hand we have by the proof of Lemma 18.18 that

$$\psi_\mu*\sigma-\psi_\mu=\mathscr{F}_G((1-\mathscr{F}_\Gamma\sigma)\mu) \quad \text{for } \sigma\in S,$$

and since $\mu_\sigma\mid G\setminus\{0\}=(1-\mathscr{F}_\Gamma\sigma)\mu$ (cf. (1)), it follows that

$$q*\sigma-q=\mu_\sigma(\{0\})\geqq 0.$$

By Lemma 18.16 this implies that q is a non-negative quadratic form on Γ, and (8) follows.

Conversely it is clear that (8) defines a continuous negative definite function with μ as Lévy measure, and it only remains to see that the quadruple (c, l, q, μ) in (8) is uniquely determined by ψ.

Since q is a quadratic form and l is a homomorphism we find for $\gamma\in\Gamma$ and $n\in\mathbb{N}$

$$q(\gamma)=\frac{q(n\gamma)}{n^2}=\frac{\psi(n\gamma)}{n^2}-\frac{il(\gamma)}{n}-\frac{c}{n^2}-\frac{1}{n^2}\int_{G\setminus\{0\}}(1-\operatorname{Re}(x,n\gamma))\,d\mu(x).$$

The integral

$$\int_{G\setminus\{0\}}\frac{1}{n^2}(1-\operatorname{Re}(x,n\gamma))\,d\mu(x),$$

converges to zero as $n\to\infty$ by the dominated convergence theorem, because

$$\lim_{n\to\infty}\frac{1}{n^2}(1-\operatorname{Re}(x,n\gamma))=0\qquad\text{for } x\in G,$$

and we have an inequality of the form (for a constant $C>0$)

$$\frac{1}{n^2}(1-\operatorname{Re}(x,n\gamma))\leqq C(1-\operatorname{Re}(x,\gamma))\qquad\text{for } n\in\mathbb{N} \text{ and } x\in G.$$

To see this inequality we put for fixed $x\in G$ and $\gamma\in\Gamma$ $(x,\gamma)=e^{i\theta}$, where $\theta\in[-\pi,\pi]$, and then we find

$$\begin{aligned}\frac{1}{n^2}(1-\operatorname{Re}(x,n\gamma))&=\frac{1}{n^2}(1-\cos n\theta)\\&=\frac{2\sin^2\frac{n\theta}{2}}{n^2}\;\frac{2\sin^2\frac{\theta}{2}}{2\sin^2\frac{\theta}{2}}\;\frac{\left(\frac{\theta}{2}\right)^2}{\left(\frac{\theta}{2}\right)^2}\\&=\left(\frac{\sin\frac{n\theta}{2}}{\frac{n\theta}{2}}\right)^2\left(\frac{\frac{\theta}{2}}{\sin\frac{\theta}{2}}\right)^2(1-\cos\theta)\\&\leqq C(1-\operatorname{Re}(x,\gamma)),\end{aligned}$$

because $\dfrac{\sin y}{y}$ is bounded away from 0 on $\left[-\dfrac{\pi}{2},\dfrac{\pi}{2}\right]$.

The equation (10) follows, and this proves that (c, l, q, μ) is uniquely determined by ψ (cf. Lemma 18.18). □

Equation (8) (and (11) below) will be referred to as the *Lévy-Khinchin formula*.

18.20. Corollary. *A continuous function $\psi\colon \Gamma \to \mathbb{R}$ is negative definite if and only if it can be written*

$$\psi(\gamma)=c+q(\gamma)+\int_{G\setminus\{0\}}(1-\operatorname{Re}(x,\gamma))\,d\mu(x) \qquad \textit{for } \gamma\in\Gamma, \tag{11}$$

where c is a non-negative constant, q is a non-negative, continuous, quadratic form on Γ and μ is a positive, symmetric measure on $G\setminus\{0\}$ such that

$$\int_{G\setminus\{0\}}(1-\operatorname{Re}(x,\gamma))\,d\mu(x)<\infty \qquad \textit{for } \gamma\in\Gamma.$$

Moreover, the triple (c, q, μ) in (11) is uniquely determined by ψ: $c=\psi(0)$, μ is the Lévy measure for ψ and q is given by the formula (10).

Proof. This is immediate by Theorem 18.19 since the Lévy measure for a real continuous negative definite function is symmetric. □

18.21. Remark. Let $(\mu_t)_{t>0}$ be a convolution semigroup on $\mathbb{R}^n$ and let $\psi\colon \mathbb{R}^n \to \mathbb{C}$ be the associated continuous negative definite function. The Lévy measure for $(\mu_t)_{t>0}$ is the measure

$$\frac{1+\|x\|^2}{\|x\|^2}\,d\mu(x),$$

where μ is the measure in the representation (2) of ψ in Theorem 10.8.

This may be seen in the same way as Proposition 18.7 was proved.

18.22. Exercise. Let φ be a real continuous positive definite function on Γ and let $k\in[\varphi(0), \infty[$. Find the uniquely determined triple (c, q, μ) in the representation (11) of the real continuous negative definite function $\psi=k-\varphi$ on Γ.

18.23. Exercise. The Lévy measure for the real continuous negative definite function $y\mapsto\|y\|^\alpha$ for $\alpha\in]0, 2[$ on $\mathbb{R}^n$ $(n\geqq 1)$ is given by

$$\mu=\frac{\alpha 2^{\alpha-1}\Gamma\left(\frac{\alpha+n}{2}\right)}{\pi^{n/2}\Gamma\left(1-\frac{\alpha}{2}\right)}\|x\|^{-\alpha-n}dx,$$

and it follows that

$$\|y\|^\alpha=\frac{\alpha 2^{\alpha-1}\Gamma\left(\frac{\alpha+n}{2}\right)}{\pi^{n/2}\Gamma\left(1-\frac{\alpha}{2}\right)}\int_{\mathbb{R}^n\setminus\{0\}}\frac{1-\cos(x,y)}{\|x\|^{\alpha+n}}\,dx \qquad \text{for } y\in\mathbb{R}^n.$$

A differential operator P on $\mathbb{R}^n$ is a "local operator" in the sense that the value at the point $x\in\mathbb{R}^n$ of the function Pf, where f is a C^∞-function on $\mathbb{R}^n$, only depends on the values of f in a neighbourhood of x, or more precisely

$$\operatorname{supp}(Pf) \subseteq \operatorname{supp}(f),$$

for all C^∞-functions f on $\mathbb{R}^n$.

We shall now study convolution semigroups on G with the property that the infinitesimal generator $(A_0, D(A_0))$ for the induced Feller semigroup (cf. 12.9) is a local operator.

18.24. Definition. A convolution semigroup $(\mu_t)_{t>0}$ on G is said to be of *local type* if for all $f \in D(A_0)$

$$\operatorname{supp}(A_0 f) \subseteq \operatorname{supp}(f).$$

If $(\mu_t)_{t>0}$ is a convolution semigroup on G of local type, then we have $A_0 f(x) = A_0 g(x)$ for all $f, g \in D(A_0)$ such that $f = g$ in some neighbourhood of x, and this property "extends" to certain functions outside $D(A_0)$.

18.25. Lemma. *Let $(\mu_t)_{t>0}$ be a convolution semigroup on G of local type and let $f \in C_b(G)$ and $g \in C(G)$ be such that*

$$g = \lim_{t \to 0} \frac{1}{t} (\mu_t * f - f) \text{ uniformly on compact subsets of } G.$$

Then

$$\operatorname{supp}(g) \subseteq \operatorname{supp}(f).$$

Proof. Let $V \subseteq G$ be an open set such that $f = 0$ in V. We shall see that $g = 0$ in V. For all $\varphi \in C_c(G)$ with $\operatorname{supp}(\varphi) \subseteq V$ and $\check{\varphi} \in D(A_0) \cap D(A_1)$ we find

$$\begin{aligned} \int g(x)\varphi(x)\,dx = \langle g, \varphi \rangle &= \lim_{t \to 0} \langle \frac{1}{t}(\mu_t * f - f), \varphi \rangle \\ &= \lim_{t \to 0} \langle f, \frac{1}{t}(\mu_t * \check{\varphi} - \check{\varphi})^\vee \rangle \\ &= \langle f, (A_1 \check{\varphi})^\vee \rangle = \langle f, (A_0 \check{\varphi})^\vee \rangle = 0, \end{aligned}$$

and it follows by Proposition 18.9 that $g = 0$ a.e. in V, which implies that $g = 0$ in V since g is continuous. □

18.26. Exercise. A convolution semigroup $(\mu_t)_{t>0}$ on G is of local type if and only if

$$\operatorname{supp}(A_p f) \subseteq \operatorname{supp}(f) \quad \text{for all } f \in D(A_p),$$

where $(A_p, D(A_p))$ is the infinitesimal generator for the induced semigroup on $L^p(G)$ for $p \in [1, \infty[$.

18.27. Theorem. *Let $(\mu_t)_{t>0}$ be a convolution semigroup on G with associated continuous, negative definite function ψ on Γ and with Lévy measure μ. The following conditions are equivalent:*

(i) $(\mu_t)_{t>0}$ *is of local type.*

(ii) $\lim_{t\to 0} \frac{1}{t} \mu_t(\complement W)=0$ *for all open neighbourhoods W of 0.*

(iii) $\mu=0$.

(iv) $\psi(\gamma)=c+i\,l(\gamma)+q(\gamma)$ *for $\gamma\in\Gamma$, where $c\geqq 0$, $l\colon\Gamma\to\mathbb{R}$ is a continuous homomorphism and $q\colon\Gamma\to\mathbb{R}$ is a non-negative, continuous quadratic form.*

Proof. (i) $\Rightarrow$ (ii). Let $W\subseteq G$ be an open neighbourhood of 0 and choose open relatively compact neighbourhoods U and V of 0 such that

$$\bar{U}\subseteq V\subseteq -W.$$

There exists by Corollary 18.10 a function $\varphi'\in D(A_0)$ with the properties

$$0\leqq\varphi'\leqq 1, \quad \varphi'=1 \text{ in } U \quad\text{and}\quad \varphi'=0 \text{ in } \complement V,$$

and the function $\varphi=1-\varphi'$ then satisfies

$$0\leqq\varphi\leqq 1 \quad\text{and}\quad 1_{\complement W}\leqq\check{\varphi}.$$

By Corollary 8.6 we have

$$\lim_{t\to 0}\frac{1}{t}(\mu_t*1-1)=\lim_{t\to 0}\frac{1}{t}(e^{-t\psi(0)}-1)=-\psi(0),$$

which shows that

$$\lim_{t\to 0}\frac{1}{t}(\mu_t*(1-\varphi')-(1-\varphi'))=-\psi(0)-A_0\varphi'$$

uniformly on G, hence by Lemma 18.25 that

$$\lim_{t\to 0}\frac{1}{t}(\mu_t*\varphi(0)-\varphi(0))=0.$$

It follows that

$$\begin{aligned}0&\leqq\liminf_{t\to 0}\frac{1}{t}\mu_t(\complement W)\leqq\limsup_{t\to 0}\frac{1}{t}\mu_t(\complement W)\\&\leqq\limsup_{t\to 0}\frac{1}{t}\langle\mu_t,\check{\varphi}\rangle=\lim_{t\to 0}\frac{1}{t}(\mu_t*\varphi(0)-\varphi(0))=0.\end{aligned}$$

(ii) ⇒ (iii). Let $f\in C_c^+(G)$ satisfy $\operatorname{supp}(f)\subseteq G\setminus\{0\}$ and choose an open neighbourhood W of 0 such that $\operatorname{supp}(f)\cap W=\emptyset$. By Proposition 18.2 we then get

$$0\leqq\langle\mu,f\rangle=\lim_{t\to 0}\langle\frac{1}{t}\mu_t,f\rangle\leqq\|f\|_\infty\lim_{t\to 0}\frac{1}{t}\mu_t(\complement W)=0.$$

(iii) ⇒ (i). For $g\in C_c^+(G)\cap D(A_0)$ we have

$$\operatorname{supp}(A_0 g)\subseteq\operatorname{supp}(g). \tag{12}$$

In fact, if $g=0$ in an open neighbourhood V of 0 then we find

$$A_0 g(0)=\lim_{t\to 0}\frac{1}{t}(\mu_t*g(0)-g(0))=\lim_{t\to 0}\langle\frac{1}{t}\mu_t,\check{g}\rangle=\langle\mu,\check{g}\rangle=0,$$

and if $g=0$ in an open neighbourhood V of $x\in G$, then $\tau_{-x}g\in C_c^+(G)\cap D(A_0)$ and $\tau_{-x}g=0$ in some neighbourhood of 0, and it follows that

$$A_0 g(x)=\tau_{-x}(A_0 g)(0)=A_0(\tau_{-x}g)(0)=0,$$

thus showing (12).

It is now easy to see that $(\mu_t)_{t>0}$ is of local type. For $f\in D(A_0)$ such that $f=0$ in an open set V, we find for all $g\in C_c^+(G)$ such that $\check{g}\in D(A_0)\cap D(A_1)$

$$\begin{aligned}\int A_0 f(x)g(x)\,dx&=\lim_{t\to 0}\langle\frac{1}{t}(\mu_t*f-f),g\rangle=\lim_{t\to 0}\langle f,\frac{1}{t}(\mu_t*\check{g}-\check{g})\check{}\,\rangle\\&=\langle f,(A_1\check{g})\check{}\,\rangle=\langle f,(A_0\check{g})\check{}\,\rangle,\end{aligned}$$

and by (12) we get

$$\operatorname{supp}((A_0\check{g})\check{}\,)\subseteq\operatorname{supp}(g).$$

If furthermore $\operatorname{supp}(g)\subseteq V$ we consequently have

$$\langle A_0 f,g\rangle=\langle f,(A_0\check{g})\check{}\,\rangle=0,$$

and it follows by Proposition 18.9 that $A_0 f=0$ in V.

(iii) ⇔ (iv). This is clear by Theorem 18.19. □

Let $(\mu_t)_{t>0}$ be a transient convolution semigroup on G with potential kernel κ. By Theorem 17.4 the measure κ is a perfect kernel, and we shall now see how the Lévy measure for $(\mu_t)_{t>0}$ is related to an arbitrary fundamental family $(\sigma_V)_{V\in\check{\mathscr{V}}}$ associated with κ.

18.28. Proposition. *Let $(\sigma_V)_{V\in\check{\mathscr{V}}}$ be a fundamental family associated with κ. For $f\in D(A_0)$ we have*

$$A_0 f=\lim_{\check{\mathscr{V}}} a_V(\sigma_V-\varepsilon_0)*f\quad\text{ in } C_0(G),$$

and the Lévy measure for $(\mu_t)_{t>0}$ *is given as the vague limit*

$$\mu = \lim_{\check{\mathscr{V}}} a_V(\sigma_V \mid G \setminus \{0\}),$$

where $a_V > 0$ *for* $V \in \check{\mathscr{V}}$ *is chosen such that* $a_V(\kappa - \sigma_V * \kappa)$ *has total mass one.*

Proof. For $V \in \check{\mathscr{V}}$ we put

$$\eta_V = a_V(\kappa - \sigma_V * \kappa),$$

and then $\lim_{\check{\mathscr{V}}} \eta_V = \varepsilon_0$ in the Bernoulli topology.

For $f \in D(A_0)$ there exists $g \in C_0(G)$ such that $f = \rho_1 * g$, and by Proposition 13.3 it follows that

$$a_V(\varepsilon_0 - \sigma_V) * \rho_1 = \eta_V - \rho_1 * \eta_V,$$

and we therefore find

$$\lim_{\check{\mathscr{V}}} a_V(\varepsilon_0 - \sigma_V) * \rho_1 * g = g - \rho_1 * g \quad \text{in } C_0(G),$$

i.e.

$$\lim_{\check{\mathscr{V}}} a_V(\sigma_V - \varepsilon_0) * f = A_0 f \quad \text{in } C_0(G).$$

Applying this equation to $f \in C_c(G)$ satisfying $\check{f} \in D(A_0)$ and $\operatorname{supp}(f) \subseteq G \setminus \{0\}$ we find

$$\lim_{\check{\mathscr{V}}} \langle a_V \sigma_V, f \rangle = A_0(\check{f})(0),$$

and therefore by 18.11 that

$$\lim_{\check{\mathscr{V}}} \langle a_V \sigma_V, f \rangle = \langle \mu, f \rangle,$$

where μ is the Lévy measure for $(\mu_t)_{t>0}$. It follows by Proposition 18.9 that

$$\lim_{\check{\mathscr{V}}} a_V \sigma_V \mid G \setminus \{0\} = \mu \ \text{vaguely.} \quad \square$$

The principle of unicity of mass can be strengthened in the case of the potential kernel κ for a convolution semigroup of local type.

18.29. Proposition. *The potential kernel* κ *for a transient convolution semigroup of local type on G satisfies the principle of* "local" unicity of mass. *For* $\nu_1, \nu_2 \in D^+(\kappa)$ *and an open set* $\omega \subseteq G$ *we have the implication:*

$$\kappa * \nu_1 = \kappa * \nu_2 \ \text{in } \omega \ \Rightarrow \ \nu_1 = \nu_2 \ \text{in } \omega.$$

Proof. It is clear that the operator $I-A_0\colon D(A_0)\to C_0(G)$ is local, and it follows by Proposition 11.10 that

$$\operatorname{supp}(f)=\operatorname{supp}((I-A_0)(\rho_1 * f))\subseteq \operatorname{supp}(\rho_1 * f) \qquad \text{for } f\in C_0(G),$$

where $(\rho_\lambda)_{\lambda>0}$ is the resolvent for $(\mu_t)_{t>0}$. This gives that

$$\operatorname{supp}(f)\subseteq \operatorname{supp}(\check{\rho}_1 * f) \qquad \text{for all } f\in C_0(G).$$

Let $f\in C_0(G)$ be given such that $\check{\rho}_1 * f\in C_c(G)$ and $\operatorname{supp}(\check{\rho}_1 * f)\subseteq\omega$. Then $\operatorname{supp}(f)\subseteq\omega$ ($f\in C_c(G)$) and we find

$$\begin{aligned}\langle \nu_1, \check{\rho}_1 * f\rangle &= \langle \rho_1 * \nu_1, f\rangle = \langle \kappa *(\varepsilon_0-\rho_1)*\nu_1, f\rangle\\ &= \langle \kappa * \nu_1, f\rangle - \langle \kappa * \nu_1, \check{\rho}_1 * f\rangle\\ &= \langle \kappa * \nu_2, f\rangle - \langle \kappa * \nu_2, \check{\rho}_1 * f\rangle\\ &= \langle \nu_2, \check{\rho}_1 * f\rangle.\end{aligned}$$

This shows (cf. Proposition 18.9) that $\nu_1=\nu_2$ in ω. □

18.30. Theorem. *Let $(\mu_t)_{t>0}$ be a transient convolution semigroup on G with potential kernel κ. The following four conditions are equivalent:*

(i) *The convolution semigroup $(\mu_t)_{t>0}$ is of local type.*

(ii) *For every fundamental family $(\sigma_V)_{V\in\check{\mathscr{V}}}$ associated with κ we have $\operatorname{supp}(\sigma_V)\subseteq V$ for all $V\in\check{\mathscr{V}}$.*

(iii) *There exists a fundamental family $(\sigma_V)_{V\in\check{\mathscr{V}}}$ associated with κ such that $\operatorname{supp}(\sigma_V)\subseteq V$ for all $V\in\check{\mathscr{V}}$.*

(iv) *For every measure $\nu\in D^+(\kappa)$, every open set $\omega\subseteq G$ such that $\nu(\omega)=0$ and for every κ-balayaged measure ν' of ν on ω we have*

$$\operatorname{supp}(\nu')\subseteq\omega^*.$$

Proof. (i) ⇒ (ii). Since $\kappa*\sigma_V=\kappa*\varepsilon_0$ in $\complement V$, this follows from Proposition 18.29.

(ii) ⇒ (iii). This is trivial.

(iii) ⇒ (i). Since $\operatorname{supp}(\sigma_V)\subseteq V$ for all $V\in\check{\mathscr{V}}$ it follows from Proposition 18.28 that the Lévy measure μ for $(\mu_t)_{t>0}$ vanishes.

(i) ⇒ (iv). Since $\kappa*\nu'=\kappa*\nu$ in ω, $\nu=0$ in ω and $\operatorname{supp}(\nu')\subseteq\bar{\omega}$, this follows from Proposition 18.29.

(iv) ⇒ (i). Let $\check{\mathscr{K}}(0)$ denote the base of all compact neighbourhoods of 0 in G. Then $(\varepsilon_0^{\complement V})_{V\in\check{\mathscr{K}}(0)}$ is a fundamental family associated with κ (cf. 17.5), and by (iv) we have

$$\operatorname{supp}(\varepsilon_0^{\complement V})\subseteq(\complement V)^*\subseteq V$$

for all $V\in\check{\mathscr{K}}(0)$. It follows like in "(iii) ⇒ (i)" that (i) holds. □

18.31. Examples. 1) The Brownian semigroup $(\mu_t)_{t>0}$ on $\mathbb{R}^n$ is of local type. In fact, the associated continuous negative definite function is a quadratic form, and Theorem 18.27 shows that $(\mu_t)_{t>0}$ is of local type. For $n \geqq 3$, $(\mu_t)_{t>0}$ is transient and the fundamental family $(\sigma_r)_{r>0}$, cf. 17.16, associated with the Newtonian kernel has property (iii) of Theorem 18.30.

2) The heat semigroup on $\mathbb{R}^{n+1}$ is of local type because the associated continuous negative definite function is of the form (iv) of Theorem 18.27.

3) The continuous negative definite function ψ associated with a convolution semigroup $(\mu_t)_{t>0}$ on $\mathbb{R}^n$ of local type has the form

$$\psi(y) = c + i \sum_{k=1}^{n} b_k y_k + \sum_{j,k=1}^{n} a_{jk} y_j y_k,$$

where $c \geqq 0$, $b_k \in \mathbb{R}$, $k = 1, \ldots, n$ and (a_{jk}) is a real, symmetric and positive semi-definite matrix.

The differential operator

$$Af = -cf - \sum_{k=1}^{n} b_k \frac{\partial f}{\partial x_k} + \sum_{j,k=1}^{n} a_{jk} \frac{\partial^2 f}{\partial x_j \partial x_k},$$

defined for C^2-functions f with compact support is (restriction of) the infinitesimal generator for the contraction semigroups on $C_0(\mathbb{R}^n)$ or $L^p(\mathbb{R}^n)$, $1 \leqq p < \infty$, induced by $(\mu_t)_{t>0}$. (Cf. Courrège [1].)

4) The symmetric stable semigroup on $\mathbb{R}^n$ of order $\alpha \in]0, 2[$ is not of local type because the Lévy measure does not vanish, cf. Exercise 18.23. This corresponds to the fact that the potential kernel, in the transient cases, has a fundamental family $(\sigma_{r,\alpha})_{r>0}$ (cf. 17.20) for which $\operatorname{supp}(\sigma_{r,\alpha}) \nsubseteq V_r$.

5) A convolution semigroup on $\mathbb{R}$ supported by $[0, \infty[$ is of local type if and only if it has the form

$$(e^{-at} \varepsilon_{bt})_{t>0} \quad \text{for } a, b \geqq 0.$$

18.32. Remark. The above method of defining the Lévy measure is inspired by Harzallah [2], which is also the source for the above proof of the Lévy-Khinchin formula (in the real case).

Convolution semigroups of local type have been studied in Forst [1], where Proposition 18.9 and a part of Theorem 18.27 are proved.

Bibliography

Artin, E.: [1] The Gamma Function. New York: Holt 1964.

Berg, C.: [1] Sur les semi-groupes de convolution. Lecture Notes in Mathematics **404**, 1-26. Berlin: Springer 1974.
[2] On the potential operators associated with a semigroup. Studia Math. **51**, 109-111 (1974).

Berg, C., Forst, G.: [1] Non-symmetric translation invariant Dirichlet forms. Invent. Math. **21**, 199-212 (1973).
[2] A remark on the behaviour at infinity of the potential kernel. Z. Wahrscheinlichkeitstheorie verw. Geb. **31**, 141-145 (1975).

Beurling, A., Deny, J.: [1] Dirichlet spaces. Proc. Nat. Acad. Sci. U.S.A. **45**, 208-215 (1959).

Bochner, S.: [1] Harmonic analysis and the theory of probability. Berkeley and Los Angeles: University of California Press 1955.

Bourbaki, N.: [1] Intégration. Ch. 1-4. 2ᵉ éd. Paris: Hermann 1965.
[2] Intégration: Ch. 6. Paris: Hermann 1959.
[3] Intégration. Ch. 7-8. Paris: Hermann 1963.
[4] Théories spectrales. Ch. 1-2. Paris: Hermann 1967.

Brelot, M.: [1] Éléments de la théorie classique du potentiel. 4ᵉ éd. Paris: C.D.U. 1969.

Choquet, G.: [1] Deux exemples classiques de représentation intégrale. Enseignement Math. **15**, 63-75 (1969).

Choquet, G., Deny, J.: [1] Aspects linéaires de la Théorie du Potentiel II. Théorème de dualité et applications. C.R. Acad. Sci. Paris **243**, 764-767 (1956).
[2] Aspects linéaires de la Théorie du Potentiel III. Noyaux de convolution satisfaisant au principe du balayage sur tout ouvert. C.R. Acad. Sci. Paris **250**, 4260-4262 (1960).
[3] Noyaux de convolution et balayage sur tout ouvert. Lecture Notes in Mathematics **404**, 60-112. Berlin: Springer 1974.

Courrège, P.: [1] Générateur infinitesimal d'un semi-groupe de convolution sur $\mathbb{R}^n$, et formule de Lévy-Khinchine. Bull. Sci. Math. **88**, 3-30 (1964).

Deny, J.: [1] Familles fondamentales. Noyaux associés. Ann. Inst. Fourier **3**, 73-101 (1951).
[2] Les noyaux élémentaires. Séminaire de Théorie du Potentiel 4. Paris 1959-60.
[3] Sur l'équation de convolution $\mu = \mu * \sigma$. Séminaire de Théorie du Potentiel 4. Paris 1959-60.
[4] Les principes fondamentaux de la théorie du potentiel. Séminaire de Théorie du Potentiel 5. Paris 1960-61.
[5] Noyaux de convolution de Hunt et noyaux associés à une famille fondamentale. Ann. Inst. Fourier **12**, 643-667 (1962).
[6] Méthodes hilbertiennes en théorie du potentiel. Potential Theory (C.I.M.E. I Ciclo, Stresa), 121-201. Rome: Ed. Cremonese 1970.

Erdélyi, A., et al.: [1] Tables of integral transforms. Vol. 1. New York: McGraw-Hill 1954.

Faraut, J.: [1] Semi-groupes de mesures complexes et calcul symbolique sur les générateurs infinitésimaux de semi-groupes d'opérateurs. Ann. Inst. Fourier **20**^{1}, 235-301 (1970).

Feller, W.: [1] An Introduction to Probability Theory and Its Applications. Vol. II, 2 ed. New York: John Wiley & Sons 1970.

Forst, G.: [1] Convolution semigroups of local type. Math. Scand. **34**, 211–218 (1974).
[2] Familles résolvantes de mesures. Séminaire de Théorie du Potentiel 17. Paris 1973–74.

Godement, R.: [1] Introduction aux travaux de A. Selberg. Séminaire Bourbaki. Exposé 144. Paris 1957.

Harzallah, K.: [1] Fonctions opérant sur les fonctions définies-négatives. Ann. Inst. Fourier, **17**[1], 443–468 (1967).
[2] Sur une démonstration de la formule de Lévy-Khinchine. Ann. Inst. Fourier, **19**[2], 527–532 (1969).

Hazod, W.: [1] Über die Lévy-Hinčin-Formel auf lokalkompakten topologischen Gruppen. Z. Wahrscheinlichkeitstheorie verw. Geb. **25**, 301–322 (1973).

Hewitt, E., Ross, K. A.: [1] Abstract Harmonic Analysis I. Berlin: Springer 1963.
[2] Abstract Harmonic Analysis II. Berlin: Springer 1970.

Hille, E., Phillips, R. S.: [1] Functional analysis and semi-groups. Amer. Math. Soc. Coll. Publ. **31** (1957).

Hirsch, F.: [1] Sur une généralisation d'un théorème de M. Itô. C.R. Acad. Sci., Paris **271**, 1236–1238 (1970).
[2] Intégrales de résolvantes et calcul symbolique. Ann. Inst. Fourier **22**[4], 239–264 (1972).
[3] Transformation de Stieltjes et fonctions opérant sur les potentiels abstraits. Lecture Notes in Mathematics **404**, 149–163. Berlin: Springer 1974.
[4] Familles d'opérateurs potentiels. To appear in: Ann. Inst. Fourier 1975.

Hunt, G.A.: [1] Semi-groups of measures on Lie groups. Trans. Amer. Math. Soc. **81**, 264–293 (1956).
[2] Markoff processes and potentials I–III. Illinois J. Math. **1**, 44–93 and 316–369 (1957); **2**, 151–215 (1958).

Itô, M.: [1] Sur les sommes de noyaux de Dirichlet. C.R. Acad. Sci. Paris **271**, 937–940 (1970).
[2] Remarque sur la somme des résolvantes. Proc. Japan Acad. **46**, 243–245 (1970).
[3] Sur une famille sous-ordonnée au noyau de convolution de Hunt donné. Nagoya Math. J. **51**, 45–56 (1973).
[4] Sur la famille sous-ordonnée au noyau de convolution de Hunt II. Nagoya Math. J. Vol. **53**, 115–126 (1974).

Landkoff, N. S.: [1] Foundations of modern potential theory. Berlin: Springer 1971.

Lukacs, E.: [1] Characteristic functions. London: Griffin 1960.

Meyer, P.-A.: [1] Probabilités et potentiel. Paris: Hermann 1966.

Ornstein, D. S.: [1] Random Walks I. Trans. Amer. Math. Soc. **138**, 1–43 (1969).

Parthasarathy, K. R.: [1] Probability measures on metric spaces. New York: Academic Press 1967.

Parthasarathy, K. R., Ranga Rao, R., Varadhan, S. R. S.: [1] Probability distributions on locally compact abelian groups. Illinois Math. J. **7**, 337–369 (1963).

Port, S. C., Stone, C. J.: [1] Potential Theory of Random Walks on Abelian Groups. Acta Math. **122**, 19–114 (1969).
[2] Infinitely Divisible Processes and Their Potential Theory I–II. Ann. Inst. Fourier **21**[2], 157–275 (1971); **21**[4], 179–265 (1971).

Rudin, W.: [1] Fourier analysis on groups. New York: Interscience 1962.

Sato, K.: [1] Potential operators for Markov processes. Proceedings of the sixth Berkeley Symposium on mathematical statistics and probability. Vol. 3, 193–211. Berkeley and Los Angeles: University of California Press 1972.

Schoenberg, I. J.: [1] Metric spaces and positive definite functions. Trans. Amer. Math. Soc. **44**, 522–536 (1938).

Schwartz, L.: [1] Théorie des distributions. Paris: Hermann 1966.

Spitzer, F.: [1] Principles of Random Walk. Princeton: Van Nostrand 1964.

Widder, D. V.: [1] The Laplace Transform. Princeton: Princeton University Press 1946.

Yosida, K.: [1] The existence of the potential operator associated with an equicontinuous semigroup of class (C_0). Studia Math. **31**, 531–533 (1968).
[2] Functional Analysis. 3rd edition. Berlin: Springer 1971.

Zygmund, A.: [1] Trigonometric Series I. Cambridge: Cambridge University Press 1968.

Symbols

$C(G)$ 1
$C_c(G)$ 1
$C_0(G)$ 1
$C_b(G)$ 1
$CN(\Gamma)$ 39
$CP(G)$ 12
$D^+(\mu)$ 4
$\mathscr{F}, \mathscr{F}_G$ 8, 24
$\overline{\overline{\mathscr{F}}}, \overline{\overline{\mathscr{F}}}_G$ 9, 109
$\hat{G}$ 8
Γ 8
$H(\mu)$ 142
$\mathscr{H}$ 122
$\mathscr{H}(\kappa)$ 132
$L^p(G)$ 3
$\mathscr{L}$ 66
$M(G)$ 1
$M_b(G)$ 1
$M_c(G)$ 1
$M_1^+(G)$ 1
$M_p(G)$ 18
N_n 111
$N(\Gamma)$ 39
$P(G)$ 12
$\mathscr{P}$ 122
$\mathscr{P}(\kappa)$ 132
R_ξ^ω 153
S 172
$S(\mu)$ 142
$\mathscr{S}$ 127
$\mathscr{S}(\kappa)$ 132
$\mathbb{T}$ 8
$\mathscr{U}_G(K, \varepsilon)$ 8
$\mathrm{cap}(\omega)$ 156
dx 3
$d\gamma$ 9
ε_a 1
k_n 111
κ_τ 130
μ_t^f 69
$\mathrm{per}(\mu)$ 30
$\sigma(f)$ 32
$\sigma^*(\mu)$ 34
τ_a 1, 2
ω_G 3
$\dot{x}$ 30
$\dot{f}$ 30
$\dot{\mu}$ 35
$\hat{}$ 8, 10
$\check{}$ 1
$\tilde{}$ 1
$\perp$ 10

General Index

approximate unit 3
associated continuous negative definite function 50
associated convolution semigroup 50
associated measure –
  for continuous positive definite function 14
  for positive definite measure 19

balayaged measure 139
balayage principle 139
balayage principle for all open sets 139
Bernoulli topology 2
Bernstein function 61
binomial distribution 27
Bochner's theorem 14
brownian semigroup 73, 95, 111

canonical balayaged measure 154
canonical equilibrium distribution 156
canonical extension of a Bernstein function 68
canonical extension of a completely monotone function 67
canonical extension of a Laplace transform 67
canonical injection 10
canonical mapping 10
capacity 156
Cauchy distribution 27
Cauchy semigroup 73
character 8
Chung-Fuchs criterion 120
circle group 8
co-Fourier transformation 9
complete maximum principle 139
completely monotone function 61
continuity theorem of Lévy 17
continuous kernel 85
continuous semigroup of measures 56
contraction resolvent 84
contraction semigroup 76
convolution 3
convolution kernel 85
convolution operator 86
convolution semigroup 48
  see also: integrable, symmetric, transient, local type
convolution semigroup determined by a measure 51
convolution semigroup supported by $[0, \infty[$ 68

degenerate distribution 27
degenerate semigroup 73
domination principle 139
dual group 8
dual Haar measure 9
dual homomorphism 10

elementary kernel 100
elementary kernel determined by μ 100
equilibrium distribution 156
equilibrium principle 156
excessive measure 146

Feller semigroup 89
Fourier transformation 8
Fourier-Stieltjes transformation 8
Fourier transformation of positive definite measure 24
fractional power of order α 133
fundamental family 160

Gamma distribution 27
Gamma semigroup 71, 73
Gaussian semigroup 73

harmonic measure 142
harmonic part of a superharmonic measure 143
heat semigroup 74, 112
Hille-Yosida theorem 84

induced semigroup 88
infinitesimal generator 77
integrable convolution semigroup 107
invariant measure 146
invariant part of an excessive measure 148
inversion theorem 10

Laplace distribution 27
Laplace transform 66
LCA-group 1
Lévy-Khinchin formula 75, 183

Lévy measure 173
local type, convolution semigroup of 185
logarithmic convex function 123
logarithmic convex sequence 122

μ-harmonic 142
μ-superharmonic 142
multiplication operator 87

negative definite function 39
Newton kernel 112, 169
normal distribution 27

one-sided stable semigroup 71, 73
orthogonal complement 10

perfect kernel 160
period 30
periodic measure 30
periodicity group 30
periodicity group for resolvent of measures 56
periodicity group for semigroup of measures 56
Plancherel's theorem 9
Poisson distribution 27
Poisson summation formula 37
Poisson semigroup 71, 73, 119
Polya's theorem 28
Pontryagin's duality theorem 10
positive definite function 12
positive definite measure 18
positive hermitian matrix 11
potential generated by a measure 143, 147
potential kernel 98
potential kernels subordinated to 132
potential operator 77
potential part of an excessive measure 148
potential part of a superharmonic measure 143
principle, see: balayage, balayage for all
open sets, complete maximum, domination
principle of local unicity of mass 188
principle of positivity of mass 156
principle of unicity of mass 139
product group 11

quadratic form 46
quotient function 32
quotient group 10
quotient measure 35

recurrent convolution semigroup 98
reduced measure 153
representing measure for a completely
monotone function 62
resolvent equation 53, 54, 81
resolvent for a contraction semigroup 81
resolvent for a convolution semigroup 53
resolvent of measures 54
Riemann-Lebesque's lemma 9
Riesz decomposition theorem 143, 148
Riesz kernel of order α 136, 171

Schoenberg's theorem 41
semigroup, see: continuous, contraction,
convolution, translation
semigroup of measures 54
semigroup subordinated to 69
shift-bounded measure 5
Stieltjes transform 127
stable semigroup, see: one-sided, symmetric
strongly continuous contraction resolvent 84
strongly continuous contraction semigroup 76
strongly subadditive capacity 160
submarkovian operator 86
subordinated to, see: semigroup subordinated
to, potential kernel subordinated to
superharmonic measure 142
symmetric convolution semigroup 51
symmetric degenerate distribution 27
symmetric stable semigroup 73, 74, 135

tend to zero at infinity, measure 5
triangular distribution 27
transient convolution semigroup 98
transition probability 60
translation 1
translation invariant continuous kernel 85
translation invariant Feller semigroup 89
translation invariant Markov process 60
translation invariant operator 86
translation semigroup 52
translation semigroup with speed a 73

uniform distribution 27

vague topology 2

zero-resolvent 82

Ergebnisse der Mathematik und ihrer Grenzgebiete

1. Bachmann: Transfinite Zahlen
2. Miranda: Partial Differential Equations of Elliptic Type
4. Samuel: Méthodes d'algèbre abstraite en géométrie algébrique
5. Dieudonné: La géométrie des groupes classiques
7. Ostmann: Additive Zahlentheorie. 1. Teil: Allgemeine Untersuchungen
8. Wittich: Neuere Untersuchungen über eindeutige analytische Funktionen
10. Suzuki: Structure of a Group and the Structure of its Lattice of Subgroups. Second edition in preparation
11. Ostmann: Additive Zahlentheorie. 2. Teil: Spezielle Zahlenmengen
13. Segre: Some Properties of Differentiable Varieties and Transformations
14. Coxeter/Moser: Generators and Relations for Discrete Groups
15. Zeller/Beekmann: Theorie der Limitierungsverfahren
16. Cesari: Asymptotic Behavior and Stability Problems in Ordinary Differential Equations
17. Severi: Il teorema di Riemann-Roch per curve, superficie e varietà questioni collegate
18. Jenkins: Univalent Functions and Conformal Mapping
19. Boas/Buck: Polynomial Expansions of Analytic Functions
20. Bruck: A Survey of Binary Systems
21. Day: Normed Linear Spaces
23. Bergman: Integral Operators in the Theory of Linear Partial Differential Equations
25. Sikorski: Boolean Algebras
26. Künzi: Quasikonforme Abbildungen
27. Schatten: Norm Ideals of Completely Continuous Operators
30. Beckenbach/Bellman: Inequalities
31. Wolfowitz: Coding Theorems of Information Theory
32. Constantinescu/Cornea: Ideale Ränder Riemannscher Flächen
33. Conner/Floyd: Differentiable Periodic Maps
34. Mumford: Geometric Invariant Theory
35. Gabriel/Zisman: Calculus of Fractions and Homotopy Theory
36. Putnam: Commutation Properties of Hilbert Space Operators and Related Topics
37. Neumann: Varieties of Groups
38. Boas: Integrability Theorems for Trigonometric Transforms
39. Sz.-Nagy: Spektraldarstellung linearer Transformationen des Hilbertschen Raumes
40. Seligman: Modular Lie Algebras
41. Deuring: Algebren
42. Schütte: Vollständige Systeme modaler und intuitionistischer Logik
43. Smullyan: First-Order Logic
44. Dembowski: Finite Geometries
45. Linnik: Ergodic Properties of Algebraic Fields
46. Krull: Idealtheorie
47. Nachbin: Topology on Spaces of Holomorphic Mappings
48. A. Ionescu Tulcea/C. Ionescu Tulcea: Topics in the Theory of Lifting
49. Hayes/Pauc: Derivation and Martingales
50. Kahane: Séries de Fourier absolument convergentes
51. Behnke/Thullen: Theorie der Funktionen mehrerer komplexer Veränderlichen
52. Wilf: Finite Sections of Some Classical Inequalities
53. Ramis: Sous-ensembles analytiques d'une variété banachique complexe
54. Busemann: Recent Synthetic Differential Geometry
55. Walter: Differential and Integral Inequalities
56. Monna: Analyse non-archimédienne
57. Alfsen: Compact Convex Sets and Boundary Integrals
58. Greco/Salmon: Topics in m-Adic Topologies
59. López de Medrano: Involutions on Manifolds
60. Sakai: C*-Algebras and W*-Algebras
61. Zariski: Algebraic Surfaces

62. Robinson: Finiteness Conditions and Generalized Soluble Groups, Part 1
63. Robinson: Finiteness Conditions and Generalized Soluble Groups, Part 2
64. Hakim: Topos annelés et schémas relatifs
65. Browder: Surgery on Simply-Connected Manifolds
66. Pietsch: Nuclear Locally Convex Spaces
67. Dellacherie: Capacités et processus stochastiques
68. Raghunathan: Discrete Subgroups of Lie Groups
69. Rourke/Sanderson: Introduction to Piecewise-Linear Topology
70. Kobayashi: Transformation Groups in Differential Geometry
71. Tougeron: Idéaux de fonctions différentiables
72. Gihman/Skorohod: Stochastic Differential Equations
73. Milnor/Husemoller: Symmetric Bilinear Forms
74. Fossum: The Divisor Class Group of a Krull Domain
75. Springer: Jordan Algebras and Algebraic Groups
76. Wehrfritz: Infinite Linear Groups
77. Radjavi/Rosenthal: Invariant Subspaces
78. Bognár: Indefinite Inner Product Spaces
79. Skorohod: Integration in Hilbert Space
80. Bonsall/Duncan: Complete Normed Algebras
81. Crossley/Nerode: Combinatorial Functors
82. Petrov: Sums of Independent Random Variables
83. Walker: The Stone-Čech Compactification
84. Wells/Williams: Embeddings and Extensions in Analysis
85. Hsiang: Cohomology Theory of Topological Transformation Groups
86. Olevskiĭ: Fourier Series with Respect to General Orthogonal Systems